普通高等教育“十四五”规划教材

Introduction to Industrial Engineering

工业工程专业导论

主编　李　杨

扫一扫查看
全书数字资源

北　京
冶　金　工　业　出　版　社
2022

内 容 提 要

本书精选工业工程专业的核心技术及前沿内容，旨在使工业工程专业学生熟悉并掌握相关专业的英文词汇，能够熟练阅读专业英文书籍与文献，提高英文的写作能力。全书共分为13章，涵盖了工业工程专业基础与专业主干课程，主要内容包括工业工程概述、工业工程应用领域、工业工程发展历史、运筹学、系统工程、工作研究（包括方法研究与作业测定）、生产计划与控制、物流工程、人因工程、质量管理、管理信息系统、人力资源管理、智能制造。

本书可作为高等院校工业工程专业本科生与研究生的教材，也可供从事工业工程专业的有关工程技术人员和管理人员参考。

图书在版编目(CIP)数据

工业工程专业导论=Introduction to Industrial Engineering：汉文、英文/李杨主编. —北京：冶金工业出版社，2022.5

普通高等教育“十四五”规划教材

ISBN 978-7-5024-9126-0

Ⅰ.①工… Ⅱ.①李… Ⅲ.①工业工程—高等学校—教材—汉、英 Ⅳ.①TB

中国版本图书馆CIP数据核字(2022)第063096号

Introduction to Industrial Engineering 工业工程专业导论

出版发行 冶金工业出版社 **电　话** (010)64027926
地　址 北京市东城区嵩祝院北巷39号 **邮　编** 100009
网　址 www.mip1953.com **电子信箱** service@mip1953.com

责任编辑 王 颖 美术编辑 彭子赫 版式设计 郑小利
责任校对 石 静 责任印制 禹 蕊

北京虎彩文化传播有限公司印刷

2022年5月第1版，2022年5月第1次印刷

787mm×1092mm 1/16；11.5印张；276千字；173页

定价 49.00元

投稿电话 (010)64027932 投稿信箱 tougao@cnmip.com.cn
营销中心电话 (010)64044283
冶金工业出版社天猫旗舰店 yjgycbs.tmall.com

前　言

本书既是一本适合于工业工程专业本科生与研究生的导论教科书，也是为工业工程专业学生学习专业英语而编写的。其目的在于进一步巩固和提高学生的专业能力与英语水平，使其能够掌握工业工程专业的英语词汇，顺利阅读工业工程专业文献，提高工业工程专业国际化水平。

本书共13章，内容包括工业工程概述、工业工程应用领域、工业工程发展历史、运筹学、系统工程、工作研究（包括方法研究与作业测定）、生产计划与控制、物流工程、人因工程、质量管理、管理信息系统、人力资源管理、智能制造。

本书突出如下特色：

（1）专业知识覆盖面广。本书针对工业工程专业的学生介绍与该专业基础课程和专业课程有关的英语基础知识和专业知识。

（2）清晰易懂。本书附有大量注释和词汇以帮助学生自己阅读。

（3）时效性。本书针对当前专业发展趋势，着重对智能制造方面做出介绍，方便学生对专业未来发展方向的了解和把握。

本书的第1章由李杨、菲罗热·再比布拉编写，第2章由李杨、王舒宁编写，第3章由李杨、樊廷阁编写，第4章和第5章由谭惠启编写，第6章由肖楠编写，第7章由王志鹏编写，第8章由李承昆编写，第9章由李娜编写，第10章由王静怡编写，第11章由李杨、侯清源编写，第12章和第13章由李杨编写。全书由李杨统稿。

本书涉及的工业工程领域的内容比较广泛，由于编者水平所限，书中难免出现疏漏与不妥之处，敬请广大读者批评指正。

编　者

2022年1月

Contents

1 Introduction

1.1 The Definition of IE

Industrial engineering is a branch of engineering which deals with the optimization of complex processes, systems or organizations. Industrial engineers work to eliminate waste of time, money, materials, person-hours, machine time, energy and other resources that do not generate value. According to the Institute of Industrial and Systems Engineers, they create engineering processes and systems that improve quality and productivity. They also develop management systems to coordinate activities and design or improve systems for the physical distribution of goods and service. Industrial engineers conduct surveys and use some calculations to find plant locations with the best combination of raw materials and transportation. An Industrial engineer maybe employed in almost any type of industry, business or institution, from retail establishments, manufacturing plants, government offices to hospitals. Industrial engineering, from Taylor's scientific management to Toyota's lean production, has played a huge role in promoting the economic and social development of countries, and it is an important basis to the industrial civilization of a country. As China gradually becomes a world manufacturing power, industrial engineering is increasingly important.

Industrial engineering is concerned with the design, development, improvement, and implementation of integrated systems of people, money, knowledge, information, machines, energy, materials, analysis and synthesis, as well as the mathematical, physical and social sciences together with the principles and methods of engineering design to specify, predict, and evaluate the results to be obtained from such systems or processes with efficiency, quality and safety. Certain words are shown in the definition[1]:

(1) **Design**. Some industrial engineering tasks involve the creation of new facility, process or system.

(2) **System**. Most engineers design physical objects but most IEs design systems. Systems include not only physical components but also processes, rules and people. A change to one part of system may affect other parts of the system.

(3) **People**. Among all types of engineers and workers.

(4) **Machines**. An IE engineer must select the appropriate machines including computers.

(5) **Information**. Data can be used for immediate decision making and can be analyzed to make improvements to the system.

(6) **Money**. An IE engineer must weight costs and savings now against costs and savings in the future and reduce the costs as much as possible.

(7) **Goal**. Every designed system exists for some purposes. The IEs must think about different ways to accomplish that goal and select the best way.

(8) **Efficiency**. Whatever the goal of the system, the IEs usually try their best to achieve the goal quickly and with the least use of resources.

(9) **Quality**. The IE's goal is that deliver goods and services to the customer with the quality that customer wants.

(10) **Safety**. IEs have to make sure that system is designed so that workers can work safely.

1.2 The Academics Domain of IE

While industrial engineering is a longstanding engineering discipline subject to (and eligible for) professional engineering licensure in most jurisdictions, its underlying concepts overlap considerably with certain business-oriented disciplines such as operations management.

Depending on the sub-specialties involved, industrial engineering may also be known as, or overlap with, operations research, systems engineering, manufacturing engineering, production engineering, management science, management engineering, ergonomics or human factors engineering, safety engineering or others, depending on the viewpoint or motives of the user. The Figure 1-1, Figure 1-2 are the **Academics domain of IE.** Figure 1-3 shows IE & other subjects.

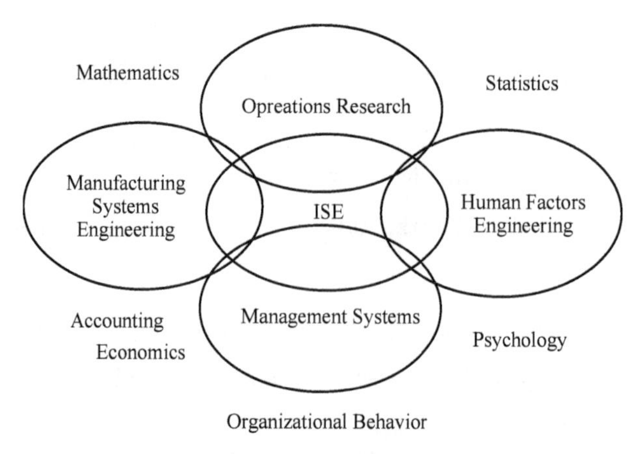

Figure 1-1 Academics domain of IE

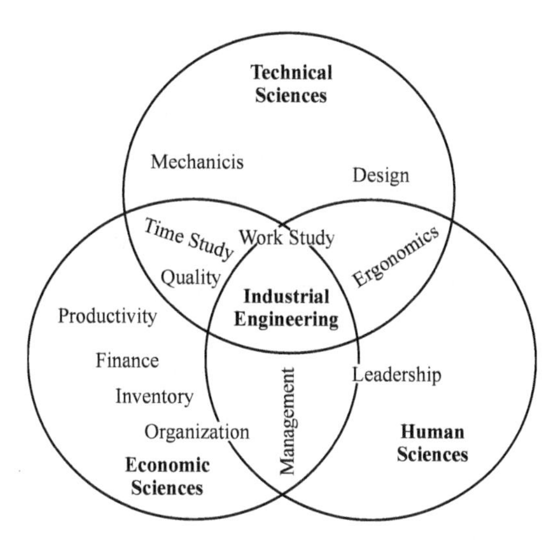

Figure 1-2 Academics domain of IE

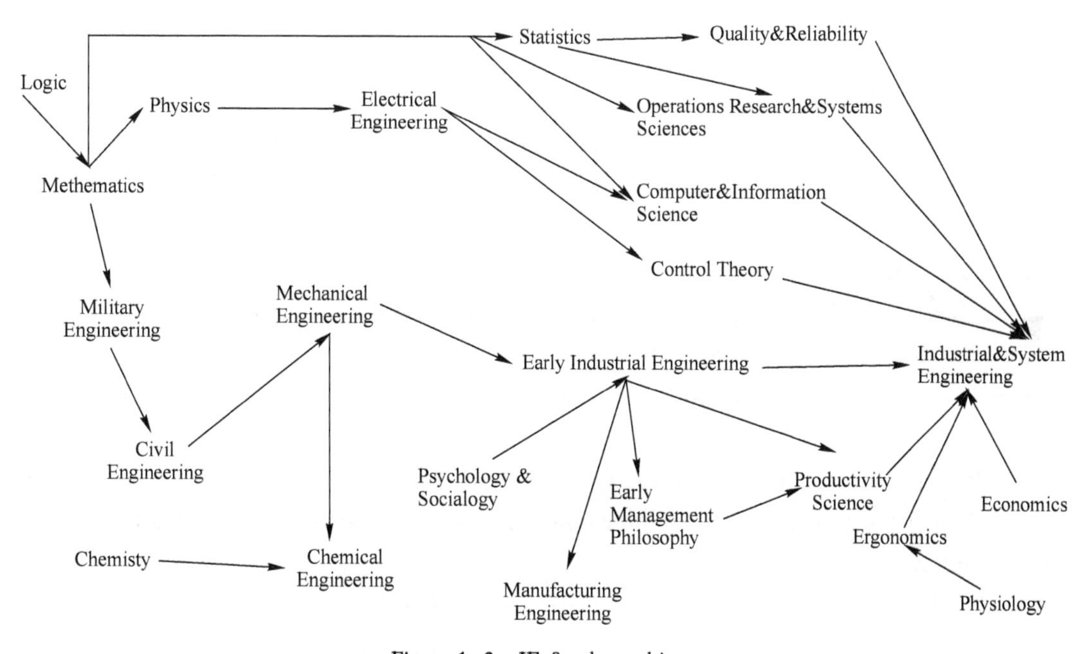

Figure 1-3 IE & other subjects

1.3 The Activities and Main Idea of IE

While originally applied to manufacturing, the use of "industrial" in "industrial engineering" can be somewhat misleading, since it has grown to encompass any methodical or quantitative approach to optimizing how a process, system or organization operates. Some engineering departments and universities have changed the term "industrial" to broader terms such as Industrial and Manufacturing Engineering, Industrial and Systems Engineering, Industrial Engineering &

Operations Research, Industrial Engineering & Management.

Activities of IE [2] :

(1) Install data processing, management information, and wage incentive systems.

(2) Develop performance standards, job evaluation, wage, and salary programs.

(3) Research new products and product applications.

(4) Improve productivity through application of technology and human factors.

(5) Select operating process and methods to do a task with proper tools and equipments.

(6) Design facilities, management systems, operating procedures.

(7) Improve planning and allocation of limited resources.

(8) Enhance plant environment and quality of workers' working life.

(9) Evaluate reliability and quality performance.

(10) Implement office systems, procedures, and policies.

(11) Analyze complex business problems by operations research.

(12) Conduct organization studies, plant location surveys, and system effectiveness studies.

(13) Study potential markets for goods and servies, raw material sources, labor supply, energy recources, financing, and taxes.

The main ideas include [2]:

(1) **Cost and efficiency**. Industrial engineering pursues the best overall benefit, requires the establishment of cost and efficiency consciousness, looking for ways to complete the work with lower cost and higher efficiency.

(2) **Problems and reforms**. IEs, should dare to reform and innovate, constantly find problems, analyze and study, find solutions, continuously improve, and pursue better.

(3) **Simplification and standardization**. The pursuit of the unity of quality and efficiency. Promote simplification, specialization, and standardization of work to achieve efficiency and cost reduction.

(4) **Overall consciousness.** The pursuit of the overall optimization of the system, according to the specific situation to choose the appropriate industrial engineering techniques, in order to obtain better overall effect.

(5) **People-oriented concept.** With the development of new architecture and new technology, people's theories, methods, and practices of industrial engineering are constantly evolving and deepening.

1.4 The Recent Research of IE

In the long-term development process of IE, new tools, new technologies and new methods are constantly adopted. From simple tools such as stopwatch at the beginning to complex tools such as Operations research, MATLAB, Python, and Genetic algoritim, IE is promoting a new journey from automation and digitalization to intelligence. IE engineers make full use of edge computing, artificial intelligence, and other technologies to realize the intellectualization of industrial manufacturing. Dynamic monitoring, intelligent data analysis, fault early warning and simulation control are constantly optimizing industrial manufacturing scenarios, processes and industrial management, greatly improving the productivity, efficiency and quality of industrial manufacturing, reducing costs, improving workers' working environment and improving safety. The realization of these goals will help promote the development of China's industry 4.0[2].

词汇

生词	音标	释义
coordinate	[kəʊˈɔːrdɪneɪt, kəʊˈɔːrdɪnət]	*n.* 坐标；（颜色协调的）配套服装 *v.* 使协调；使相配合
lean	[liːn]	*v.* 倾斜；倚靠；靠置；使斜靠 *adj.* 无脂肪的；贫乏的
synthesis	[ˈsɪnθəsɪs]	*adj.*（人工的）合成；综合体
integrated	[ˈɪntɪɡreɪtɪd]	*adj.* 集成的；综合的 *v.*（使）合并，成为一体；（使）加入
discipline	[ˈdɪsəplɪn]	*n.* 纪律；惩罚；训导，科 *v.* 惩罚，管教
jurisdiction	[ˌdʒʊrɪsˈdɪkʃn]	*n.* 司法权，审判权；管辖权
encompass	[ɪnˈkʌmpəs]	*vt.* 包含；包围，环绕

长难句

Dynamic monitoring, intelligent data analysis, fault early warning and simulation control are constantly optimizing industrial manufacturing scenarios, processes and industrial management, greatly improving the productivity, efficiency and quality of industrial manufacturing, reducing costs, improving workers' working environment and improving safety.

动态监控、智能数据分析、故障预警和仿真控制等正不断优化工业制造场景、流程和工业管理，大幅提高工业制造的生产率、效率和质量，降低成本，改善工人的工作环境并提高安全度。

Reference

[1] Bonnie B. Introduction to industrial engineering [M]. America: Mavs Open Press, 2020.
[2] 卢佩琳. 工业工程、5G 与智能制造 [J]. 工业技术创新, 2021, 8 (1): 101-107.

2 Fields of Industrial Engineering

Industrial engineering is widely used in various fields. **It was first emerged and gradually developed in the manufacturing industry in order to improve the production mode and so as to improve the production efficiency and reduce the cost better**. With the development of the times and social progress, the application of IE technology has gradually expanded to other fields such as service industry, industry, construction and so on. The various fields and topics that industrial engineers are involved with include:

Manufacturing Engineering: Although the field of IE is so extensive, manufacturing is still the most important and representative application field. First of all, the layout of the factory is very important. The area of the factory must be fully utilized. The logistics needs the shortest path, and it must be continuously improved. Secondly, quality awareness is particularly important in enterprises. In recent years, special emphasis has been placed on the design quality of products. Without high-quality design, there will be no high-quality products. The third is the time-based competitive strategy. This strategy regards time as the most valuable resource. Reduce the waste of time, so as to increase the competitiveness of products. If the enterprise's efficiency reaches two to three times that of its competitors, it will win the market.

Engineering management: Industrial engineering aims to reduce costs, improve benefits and efficiency. This goal is consistent with the goal of management. Management is a leading activity to achieve the objectives of the system, conduct command and decision-making. Industrial engineering is a technical activity engaged in planning, design, evaluation, innovation, and control. To some extent, it serves to achieve management objectives[1].

Process engineering: Process engineering is the design, operation, control, and optimization of chemical, physical, and biological processes. It mainly studies energy flow of material, involving complex chemical reactions and physical state changes. Continuity and **multivariable** are remarkable characteristics of process engineering. New technological processes emerge one after another and chemical systems are becoming larger and more complex. Modern research and development work has large investment, short cycle, high risk and fierce competition. Therefore, some of the existing engineering methods and classical computing technologies are inadequate. So, it is necessary to describe and grasp the process system from the macro as a whole.

Systems engineering: Systems engineering is an **interdisciplinary** field of engineering that focuses on how to design and manage complex engineering systems over their life cycles. With the development of the times, modern industrial engineering and system engineering are getting closer and closer. The manufacturing system oriented by industrial engineering is becoming more and more complex and considered more and more **comprehensively**. It begins to use more operation research methods, **quantitative** methods, simulation technology, computer science and so on. Therefore, modern industrial engineering can be regarded as "system engineering + traditional industrial engineering method"[2].

Software engineering: It is an interdisciplinary field of engineering that focusing on design, development, maintenance, testing, and evaluation of the software that make computers or other devices containing software work. With the rapid development of economy, the degree of informatization and intelligence of industrial products has been greatly improved. In modern industrial engineering, the degree of automation determines whether the production efficiency is efficient, Software engineering is the basis of realizing intelligence. Only by deeply studying electronic technology can industrial engineering realize modernization[3].

Safety engineering: It is an engineering discipline which assures that engineered systems provide acceptable levels of safety. **IE involves manufacturing working environment, public living space, consumer product design, man-machine interface, auxiliary equipment for the disabled, etc., which can be applied to safety engineering**. The research and development in the future should pay more attention to how to enhance the value of human nature and improve the quality of life[4].

Data science: the science of exploring, manipulating, analyzing, and visualizing data to derive useful insights and conclusions. To accurately predict the production period and cost, and accurately and **dynamically** schedule manufacturing resources in real time according to factors such as product demand changes and the current situation of materials and equipment, it is necessary to analyze manufacturing big data resources. Manufacturing big data application analysis is an important development direction of industrial engineering technology in the big data era[5].

Machine learning: the automation of learning from data using models and algorithms. With the explosive growth of the industry data, more and more attention is paid to big data. However, due to the volume, complex and fast-changing characteristics of big data, traditional machine learning algorithms for small data are not applicable. Therefore, developing machine learning algorithms for big data is a research focus[6].

Analytics and data mining: the discovery, interpretation, and extraction of patterns and insights from large quantities of data. It can help business operation, improve products and help enterprises make better decisions.

Cost engineering: practice devoted to the management of project cost, involving such activities as cost and control-estimating, which is cost control and cost forecasting, investment appraisal, and risk analysis.

Value engineering: a systematic method to improve the "value" of goods or products and services by using an examination of function. The main goal of this discipline is to improve the economic benefits of industrial enterprises and the core content is to promote the improvement of old products and the development of new products[7].

Predetermined motion time system: a technique to quantify time required for repetitive tasks. The application of the predetermined motion time system can provide an important basis for designing the working position and determining the working method[8].

Quality engineering: a way of preventing mistakes or defects in manufactured products and avoiding problems when delivering solutions or services to customers. IE technology needs to be used in quality management activities.

Project management: the process and activity of planning, organizing, motivating, and controlling resources, procedures, and protocols to achieve specific goals in scientific or daily problems. It uses specialized knowledge, skills, tools, and methods in project activities to enable the project to achieve or exceed the set needs and expectations under the limited resources.

Supply chain management: the management of the flow of goods. It includes the movement and storage of raw materials, work-in-process inventory, and finished goods from point of origin to point of consumption. The supply chain management optimize the operation of the supply chain and make the whole process of the supply chain from procurement to meeting the final customers at the least cost.

Ergonomics: the practice of designing products, systems, or processes to take proper account of the interaction between them and the people that use them. This discipline focuses on human factors and applies human scientific knowledge such as psychology, physiology, anatomy, and anthropometry to engineering technology design and operation management, especially safety design and safety management. **It focuses on improving people's work performance and preventing people's mistakes and uniformly consider the optimization of the overall performance of the human-machine-environment system under the condition of making the personnel in the system safe and comfortable as far as possible.**

Operations research: also known as management science: discipline that deals with the application of advanced analytical methods to help make better decisions. It is an emerging discipline developed in the early 1930s. Its main purpose is to provide scientific basis for managers in decision-making. It is one of the important methods to realize effective management, correct decision-making, and modern management. This discipline is applied to the interdisciplinary research of mathematics and formal science. It uses statistics, mathematical models and algorithms to find the best or approximate best solutions to complex problems[9].

Operations management: an area of management concerned with overseeing, designing, and controlling the process of production and redesigning business operations in the production of goods or services. The main objectives of operation management are quality, costly, time and flexibility, which are the fundamental source of enterprise competitiveness. Therefore, operation management plays an important role in enterprise management.

Job design: the specification of contents, methods, and relationship of jobs in order to satisfy technological and organizational requirements as well as the social and personal requirements of the job holder. Job design refers to the process of defining the tasks, responsibilities, powers of each post and the relationship with other posts in the organization according to the needs of the organization and considering the needs of individuals. It combines the work content, work qualification and remuneration in order to meet the needs of employees and organizations. Whether the post design is appropriate or not has a significant impact on stimulating employees' enthusiasm, enhancing employees' satisfaction and improving job performance[10].

Financial engineering: the application of technical methods, especially from mathematical finance and computational finance, in the practice of finance. Financial engineering focuses on the pricing and practical application of derivative financial products. It is most concerned about how to use innovative financial instruments to more effectively distribute and redistribute various economic risks faced by individuals, so as to optimize their risk or return. More specifically, financial engineering is to fit the expected rate of return of a financial securities asset to the risk-free rate of return under the assumption of risk neutrality. Instead, it tends to be a risk averse investment type.

Industrial plant configuration: sizing of necessary infrastructure used in support and maintenance of a given facility. **It refers to the specific planning and design of the system, including site selection, plant layout, logistics analysis, material handling methods and equipment selection so as to make each production factor and subsystem reasonably configured according to the requirements of industrial engineering and form a more productive integrated system.** It is the key link for industrial engineering to realize the overall optimization of the system and improve the overall benefit.

Facility management: an interdisciplinary field devoted to the coordination of space, infrastructure, people, and organization. Facility management comprehensively utilizes the theories of management science, construction science, behavior science, engineering technology and other disciplines, combines people, space and process, effectively plans and controls human work and living environment, maintains high-quality activity space, improves investment efficiency, and meets the requirements of strategic objectives and business plans of various enterprises, institutions and government departments.

Engineering design process: formulation of a plan to help an engineer build a product with a specified performance goal. Engineering design process is not a simple product modeling activity, but a composite process including technical, social, and cognitive processes. Only by organizing the design process reasonably and effectively can enterprises develop products that really meet the market demand.

Logistics: the management of the flow of goods between the point of origin and the point of consumption in order to meet some requirements, of customers or corporations. Logistics engineering is the combination of management engineering and technical engineering. It is closely related to transportation engineering, management science and engineering, system engineering, computer technology, environmental engineering, mechanical engineering, industrial engineering, architecture, and civil engineering.

Accounting: the measurement, processing, and communication of financial information about economic entities. Engineering accounting and cost control play an important role in engineering, which are closely related to the economic benefits of the project. If relevant departments want to effectively manage the project, they should reasonably carry out project accounting and control the project cost on the basis of ensuring the project quality.

Capital projects: the management of activities in capital projects involves the flow of resources, or inputs, as they are transformed into outputs. Many of the tools and principles of industrial engineering can be applied to the configuration of work activities within a project. The application of industrial engineering and operations management concepts and techniques to the execution of projects has been thus referred to as Project Production Management.

Traditionally, a major aspect of industrial engineering was **planning the layouts of factories** and designing assembly lines and other manufacturing paradigms. And now, in **lean manufacturing** systems, industrial engineers work to eliminate wastes of time, money, materials, energy, and other resources.

Examples of where industrial engineering might be used include flow process charting, process mapping, designing an assembly workstation, strategizing for various operational logistics, consulting as an efficiency expert, developing a new financial algorithm or loan system for a bank, streamlining operation and emergency room location or usage in a hospital, planning complex distribution schemes for materials or products (referred to as supply-chain management), and shortening lines (or queues) at a bank, hospital, or a theme park.

Modern industrial engineers typically use predetermined motion time system, computer simulation (especially discrete event simulation), along with extensive mathematical tools for modeling, such as mathematical optimization and queueing theory, and computational methods for system analysis, evaluation, and optimization. Industrial engineers also use the tools of data science and machine learning in their work owing to the strong relatedness of these disciplines with the field and the similar technical background required of industrial engineers (including a strong foundation in probability theory, linear algebra, and statistics, as well as having coding skills).

词汇

生词	音标	释义
multivariable	[ˌmʌltiˈveəriəbəl]	*adj.* 多变量的；多变数的
interdisciplinary	[ˌɪntəˌdɪsˈplɪnərɪ]	*adj.* 多学科的；跨学科的
comprehensively	[ˌkɒmprɪˈhensɪvlɪ]	*adv.* 完全地；彻底地
quantitative	[ˈkwɒntɪtətɪv]	*adj.* 定量性的；量化的；数量的
dynamically	[daɪˈnæmɪklɪ]	*adv.* 动态地；充满活力地；不断变化的
appraisal	[əˈpreɪzl]	*n.* 评价；估价；估计；鉴定；(上司对雇员的）工作鉴定会；工作表现评估

长难句

It was first emerged and gradually developed in the manufacturing industry in order to improve the production mode and so as to improve the production efficiency and reduce the cost better.

它最初是在制造业中出现并逐渐发展起来的，目的是改进生产模式，从而提高生产效率，更好地降低成本。

It is an interdisciplinary field of engineering that focusing on design, development, maintenance, testing, and evaluation of the software that make computers or other devices containing software work.

它是一个跨学科的工程领域，专注于软件的设计、开发、维护、测试和评估进而让计算机与软件设备正常运转。

IE involves manufacturing working environment, public living space, consumer product design, man-machine interface, auxiliary equipment for the disabled, etc. , which can be applied to safety engineering.

IE 涉及制造工作环境、公共生活空间、消费品设计、人机界面、残疾人辅助设备等，这些都可应用于安全工程。

It focus on improving people′s work performance and preventing people′s mistakes and uniformly consider the optimization of the overall performance of the human-machine-environment system under the condition of making the personnel in the system safe and comfortable as far as possible.

其重点在于提高人们的工作绩效和预防人的失误，并在尽可能地使系统中的人员安全舒适的条件下，统一考虑人机环境系统整体性能的优化。

It refers to the specific planning and design of the system, including site selection, plant layout, logistics analysis, material handling methods and equipment selection so as to make each production factor and subsystem reasonably configured according to the requirements of industrial engineering and form a more productive integrated system.

（工业设备配置）指系统的具体规划设计，包括选址、厂房布局、物流分析、物料搬运方式、设备选型等，使各生产要素和子系统按照工业工程的要求合理配置，形成更具生产性的集成系统。

Reference

[1] 邢慧贤，赵德安，亢跃华．现代工业工程（IE）技术在建筑工程管理中的应用［J］. 四川建筑，2008（4）：227-228.

[2] 卢佩琳．工业工程、5G 与智能制造［J］. 工业技术创新，2021，8（1）：35-40.

[3] 王洋．电子技术在工业工程中的应用探析［J］. 科学技术创新，2018（5）：167-168.

[4] 王国明，简祯富．台湾工业工程理论及应用发展回顾［J］. 工业工程与管理，2003（3）：1-5，23.

[5] 吴小东，赵晶英．工业工程专业制造大数据应用分析方向课程设置［J］. 装备制造技术，2018（5）：246-248.

[6] 孙志军，薛磊，许阳明，等．深度学习研究综述［J］. 计算机应用研究，2012，29（8）：2806-2810.

[7] 何盛明．财经大辞典［M］. 北京：中国财政经济出版社，1990.

[8] 陈龙华，任志明．PTS 管理方法在冶金成台设备制造生产管理上的应用［J］. 现代冶金，2000，28（3）:11-14.

[9] 蒋智凯．浅谈运筹学教学［J］. 重庆科技学院学报（社会科学版），2010（24）：176-177.

[10] 林崇德．心理学大辞典［M］. 上海：上海教育出版社，2003.

3 History of Industrial Engineering

3.1 The Preparatory Stage of Industrial Engineering

There is a general consensus among historians that the roots of the industrial engineering profession date back to the Industrial Revolution. The technologies that helped mechanize traditional manual operations in the textile industry including the flying shuttle, the spinning jenny, and **perhaps most importantly the steam engine generated economies of scale that made mass production in centralized locations attractive for the first time**. The concept of the production system had its genesis in the factories created by these innovations[1]. It has also been suggested that perhaps Leonardo da Vinci was the first Industrial Engineer, because there is evidence that he applied science to the analysis of human work, by examining the rate at which a man could shovel dirt around the year 1500. Others also state that the IE profession grew from Charles Babbage's study of factory operations and specifically his work on the manufacture of straight pins in 1832. **However, it has been generally argued that these early efforts, while valuable, were merely observational and did not attempt to engineer the jobs studied or increase overall output.**

3.2 The First Stage in the Development of Industrial Engineering

Most analysts agree that the IE profession grew out of application of science to the design of work and production systems that started in the 1880's. The pioneers who led that effort were Frederick W. Taylor. Frederick Taylor (1856 ~ 1915) is generally credited as being the father of the Industrial Engineering discipline. Management and the Principles of Scientific Management which were published in the early 1900s, were the beginning of Industrial Engineering. Improvements in work efficiency under his methods was based on improving work methods, developing of work standards, and reduction in time required to carry out the work. With an abiding faith in the scientific method, Taylor's contribution to "Time Study" sought a high level of precision and predictability for manual tasks[1].

Frank B. Gilbreth (1868 ~ 1924) and Lillian Gilbreth (1878 ~ 1972) were the other cornerstone of the Industrial Engineering movement whose jobs were housed at Purdue University School of Industrial Engineering. They extended Taylor's work considerably. By classifying motions as "grasp" "transport" and so on, they are able to measure the time of workers performing their tasks. This development permitted analysts to design jobs without knowledge of the time required to do a job. These developments were the beginning of a much broader field known as human factors or ergonomics[1].

Henry Laurence Gantt (1861 ~ 1919) developed the Gantt chart which was a significant contribution in which it provided a systematic graphical procedure for preplanning and scheduling work activities, reviewing progress, and updating the schedule. The Gantt chart is still used today.[1]

In 1917, F. W. Harris studied economic order quantity theory.

In 1922, Marken wrote Budget Control, and in 1924, Accounting Management.

3.3 The Second Stage in the Development of Industrial Engineering

In 1911, the Department of Mechanical Engineering of Purdue University first set up the industrial Engineering elective course.

In 1918, Pennsylvania State University established the Department of Industrial Engineering.

In 1920, the United States established the American Society of Industrial Engineers. There were jobs in factories that specialized in IE.

In 1924~1931, W. A. Shewhart initiated "statistical quality management".

From 1924 to 1933, G. F. Mayo pioneered the theory of interpersonal relations through the Hawthorne Experiment. J. Fish pioneered the "engineering economy".

3.4 The Third Stage in the Development of Industrial Engineering

In the mid 1940s, the United Kingdom and the United States published information about operational research achievements, which immediately attracted the attention of IE workers. OR is a system including mathematical programming, optimization theory, queuing theory, storage theory, game theory and other theories and methods, which can be used to describe, analyze and design a variety of different types of operating systems.

Since the third stage, the major research and analysis direction of industrial engineering has changed from qualitative and empirical analysis to quantitative analysis[2].

3.5 The Fourth Stage in the Development of Industrial Engineering

In the 1950s and 1960s, systems science (SS) made great progress. **System Engineering (SE), which inherits the SS idea and contains the knowledge of natural science and social science and claims to be based on OR theory and attaches great importance to Engineering application, stands out.** SE attaches great importance to the cultivation of system philosophy and training of system analysis methods and contains abundant knowledge of natural science and social science, which is exactly a "master" discipline required by IE.

In this period, industrial engineering is not only the "patent" of developed countries in Europe and America, but also has been successfully introduced into the Asia-pacific region. The most typical and successful example is the Japanese. In the post-war economic recovery period, they successfully introduced industrial engineering from American management thinking and technical means into all walks of life in Japan, and carried out Japanese-style absorption and transformation, creating Toyota Production System (TPS), Total Quality Control (TQC), etc. And quality became an important research goal of IE. Many important methods were invented during this period, such as "Six Sigma" management and total quality management. These methods are still very effective until today[2].

3.6 The Fifth Stage in the Development of Industrial Engineering

In the 1970s, Japan introduced JIT (Just In Time), It was summarized as Lean Production and Total Productive Maintenance (PTM) in the 1990s in the United States. Singapore, Thailand, and other countries have established research, education, development and application systems of IE to strengthen the development and application of industrial engineering.

After the 1990s, industrial engineering has developed by leaps and bounds. First, on the basis of the original MRP (Material Require Planning) closed-loop system, it has developed into MRP system, and then into ERP (Enterprise Resource Planning) system, which has truly added information management and realized the full use of global resources.

Since the 1970s, especially in the last 10 years, the development of IE has seen some new trends:

(1) Early IE was to improve the manufacturing site operation efficiency and improve production management; modern IE is for the whole process of enterprise management.

(2) Early IE went it alone; modern IE has become an effective tool to provide management integration infrastructure for enterprise CIMS, and then for enterprise development to LAF enterprise.

(3) Early IE was only used in manufacturing; modern IE has spread to industries as diverse as transportation, construction, service, and administration (such as the planning and design of the national health care system)[2].

3.7 The Disciplinary History of Industrial Engineering

There are a number of significant events and developments that have occurred in the evolution of industrial engineering in the following figure. Now, let's introduce the chart. Firstly, at the top of the chart shows four labels, which represent important development stages of industrial engineering. Secondly, at the middle of the chart displays important disciplines in industrial engineering. Finally, at the bottom of the chart reveals important historical points in industrial engineering.

As shown in Figure 3-1, industrial engineering is divided into four important periods. The period

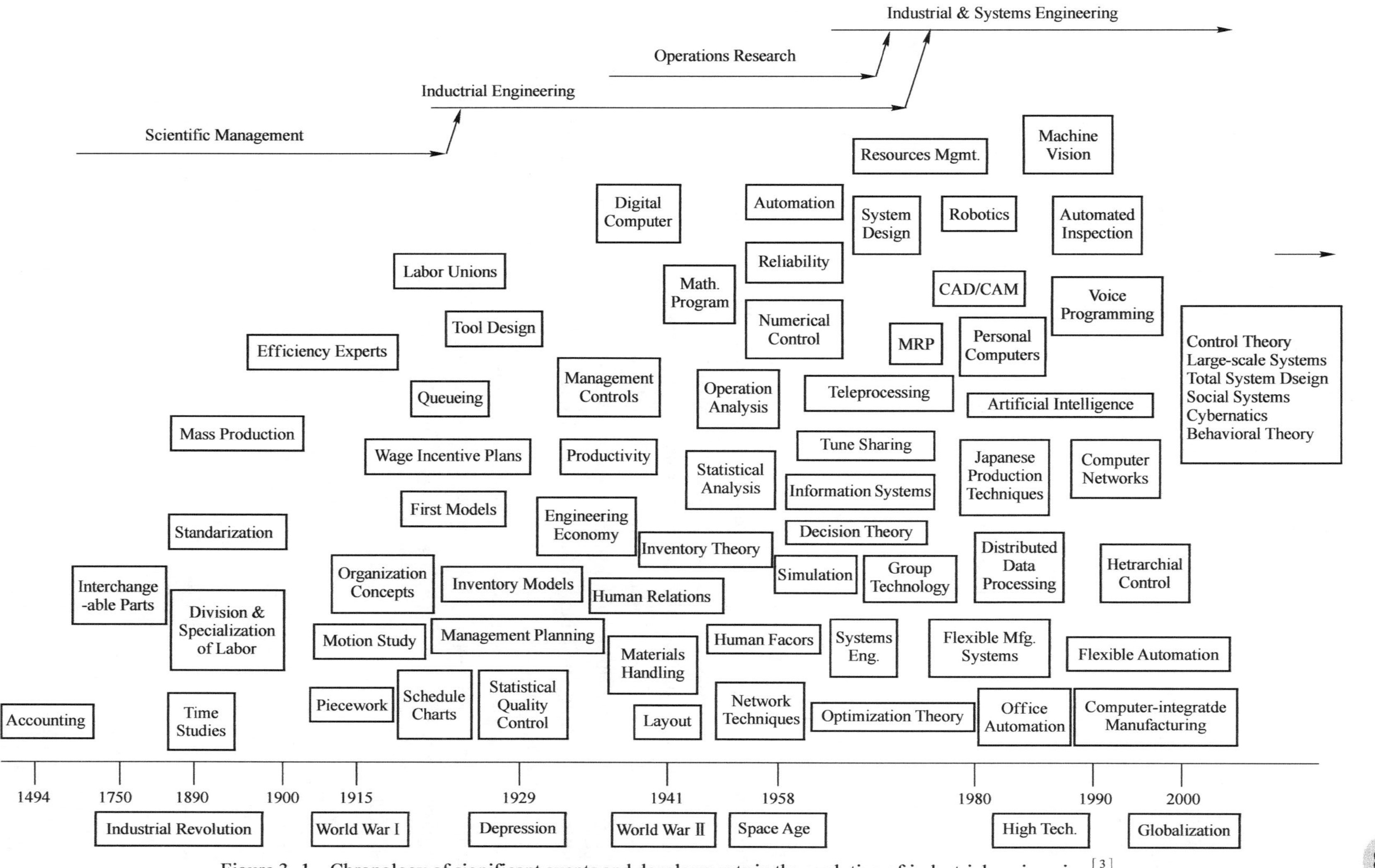

Figure 3–1 Chronology of significant events and developments in the evolution of industrial engineering[3]

from about the beginning of the nineteenth century to about the 1930s was the period of scientific management. The period from the 1920s up to the present is the period of industrial engineering. The third period is the operational research period, which lasted from the 1940s to the 1970s (in fact, even in modern times, operational research is still widely applied). The last period is the industrial and system engineering period, which lasts from 1970s to the present. It should be noted here that the relationship between one period and another does not end and the other begins, but there is a period of transition. It can be seen that the disciplines related to industrial engineering involve a wide range of fields and contents and have developed rapidly in modern times. There is overlap and correlation between the periods.

3.8 Development of Industrial Engineering in China

In the past decade, the industrial engineering profession in China has experienced two stages of development.

The first stage (1999 ~ 2000): introduction of basic concepts, theories and methods of industrial engineering.

The second stage (2000 ~ present): integrate IE's important technologies and methods with Chinese manufacturing enterprises and pay attention to the localization of IE. His research focuses on human factors Engineering, operations research, systems.

Application status of Industrial Engineering in China approved by the former State Education Commission in July 1993, Tianjin University, Xi'an Jiaotong University, the first batch of pilot industrial engineering major and recruit undergraduate students, thus creating a precedent of industrial engineering discipline in China. Later, Chongqing University and other universities also set up industrial engineering major. Up to now, the national industrial engineering professional universities have reached more than one hundred. In recent years, China's industrial engineering discipline development rapidly, industrial engineering ideas in all walks of life have been recognized, more and more fields began to apply industrial engineering.

词汇

生词	音标	释义
flying shuttle		飞梭
the spinning jenny		珍妮纺织机
shovel	[ˈʃʌvl]	*v.* 铲
manual	[ˈmænjuəl]	*adj.* 手工的，体力的
ergonomics	[ˌɜːgəˈnɒmɪks]	*n.* 工效学，人类工程学
economic order quantity		经济订货批量
post-war		战后的
all walks of life		各行各业
JIT (Just In Time)		准时制生产

lean production 精益生产
closed-loop system 闭环系统

长难句

Perhaps most importantly the steam engine generated economies of scale that made mass production in centralized locations attractive for the first time.

也许最重要的是蒸汽机产生了规模经济，使集中大规模生产第一次具有吸引力。

However, it has been generally argued that these early efforts, while valuable, were merely observational and did not attempt to engineer the jobs studied or increase overall output.

然而，人们普遍认为，这些早期的努力虽然有价值，但只是观察性的，并没有试图设计研究的工作或增加总体产出。

System Engineering (SE), which inherits the System Science (SS) idea and contains the knowledge of natural science and social science and claims to be based on OR theory and attaches great importance to Engineering application, stands out.

继承了系统科学思想，包含了自然科学和社会科学的知识，主张以运筹学理论为基础，重视工程应用的系统工程闪亮登场。

Reference

[1] Maynard H B, et al. Maynard's industrial engineering handbook [M]. New York: McGraw Hill Professional 5th Edition, 2001: 14-16.
[2] 赵涛，齐二石．工业工程发展历程与趋势浅析 [J]. 机械设计，1997 (9): 12-20.
[3] Turner W C, et al. Introduction to industrial and systems engineering [M]. 北京：清华大学出版社，2002.

4 Operations Research

4.1 Introduction to Operations Research

4.1.1 Origin and Characteristics of Operations Research

The concept of operations research arose during World War Ⅱ by military planners. After the war, the techniques used in their operations research were applied to addressing problems in business, the government and society.

There are three primary characteristics of all operations research efforts:

(1) **Optimization.** The purpose of operations research is to achieve the best performance under the given circumstances. Optimization also involves comparing and narrowing down potential options.

(2) **Simulation.** This involves building models or replications in order to try out and test solutions before applying them.

(3) **Probability and statistics.** This includes using mathematical algorithms and data to uncover helpful insights and risks, make reliable predictions and test possible solutions.

4.1.2 The Process of Operations Research

Operations research (OR) is an analytical method of problem-solving and decision-making that is useful in the management of organizations. In operations research, problems are broken down into basic components and then solved in defined steps by mathematical analysis.

The process of operations research can be broadly broken down into the following steps:

(1) Identifying a problem that needs to be solved.

(2) Constructing a model around the problem that resembles the real world and variables.

(3) Using the model to derive solutions to the problem.

(4) Testing each solution on the model and analyzing its success.

(5) Implementing the solution to the actual problem.

(6) Disciplines that are similar to, or overlap with, operations research include statistical analysis, management science, **game theory**, **optimization theory**, **artificial intelligence** and network analysis. All of these techniques have the goal of solving complex problems and improving **quantitative** decisions.

4. 1. 3 Importance of Operations Research

The field of operations research provides a more powerful approach to decision making than ordinary software and data analytics tools. Employing operations research professionals can help companies achieve more complete datasets, consider all available options, predict all possible outcomes and estimate risk. Additionally, operations research can be tailored to specific business processes or use cases to determine which techniques are most appropriate to solve the problem.

4. 2 The Main Content of Operations Research

Operations research should generally include linear programming, nonlinear programming, integer programming, dynamic programming, multi-objective programming, network analysis, queuing theory, game theory, decision theory, storage theory, reliability theory, model theory, input-output analysis, etc.

The five parts of linear programming, non-linear programming, integer programming, dynamic programming, and multi-objective programming are collectively called programming theory, and they mainly solve two aspects of problems. One aspect of the problem is how to maximize the benefits of given humans, materials and financial resources; the other aspect is how to use the least humans, materials and financial resources to complete a given task.

Network analysis is mainly to study and solve problems such as the shortest path problem, the minimum connection problem, the minimum cost flow problem, and the optimal allocation problem in production organization and planning management. Especially when designing and arranging large and complex projects, network technology is an important tool.

Queuing is common in daily life, such as machines waiting for repairs, ships waiting for loading and unloading, and customers waiting for service. They have a common problem, which is that the waiting time is long, which will affect the completion of the production task, or the customer will leave automatically, which will affect the economic benefits; if you increase the number of repairmen, loading docks and service desks, it will certainly solve the problem of excessive waiting time, but will suffer the loss of repairman, dock and service desk idleness. The proper solution of such problems is the task of antithesis. Multiple service desks in series queuing system is shown in Figure 4-1.

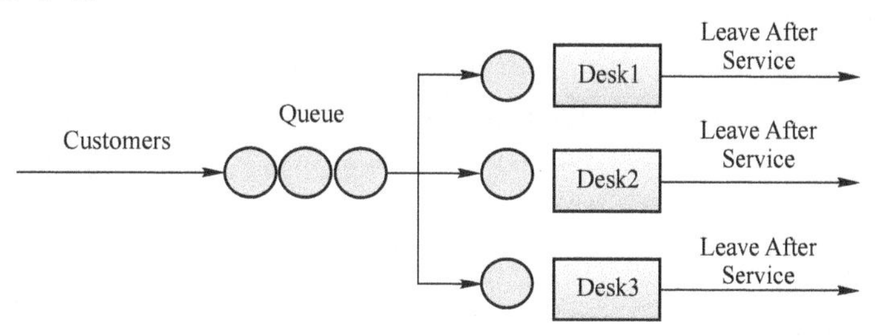

Figure 4-1 Multiple service desks in series queuing system

Game theory is to study how the parties with severe conflicts can work out strategies that are beneficial to themselves and defeat their opponents. The reason why people are hesitating is that when people set out to achieve a certain expected goal, they have a variety of situations in front of them, and there are a variety of actions to choose from. How the decision-maker chooses an optimal plan to achieve his expected goal is the research task of decision theory.

In the process of production and consumption, people must reserve a certain amount of raw materials, semi-finished products, or commodities. Less storage will cause losses due to downtime or lost sales opportunities, and too much storage will cause a backlog of funds and loss of raw materials and commodities. Therefore, how to determine a reasonable storage volume, purchase batch size and purchase cycle is very important. This is the problem that the storage theory must solve.

For a complex system and equipment, it is often composed of thousands of working units or parts. The quality of these units or parts will directly affect the stability and reliability of the system or equipment. Studying how to ensure the reliability of the system or equipment is the task of reliability theory. The things that people encounter in production practice and social practice are often very complicated. To understand the changing laws of these things, you must first properly describe the changing process of these things, which is the so-called establishment of models, and then you can Model research to understand the changing laws of things. Model theory is to study the basic skills of building models theoretically and methodically. Input-output analysis is to formulate the development plan of each department by studying the comprehensive balance principle that must be observed by the input and output of multiple departments, so as to control and adjust the national economy at a macro level to achieve a coordinated and reasonable development of the national economy.

4.3 Models and Applications of Operations Research

Operation Research model is an idealized representation of the real-life situation and represents one or more aspects of reality. Examples of operation research models are a map, activity charts balance sheets, PERT network, **break-even** equation, economic ordering quantity equation, etc. Objective of the model is to provide a means for analyzing the behavior of the system for improving its performance.

Classification of Models:

Models can be classified on the basis of following factors:

(1) By degree of Abstraction: 1) Mathematical models; 2) Language models.

(2) By Function: 1) Descriptive models; 2) Predictive models; 3) Normative models for repetitive problems.

(3) By Structure: 1) Physical models; 2) Analogue (graphical) models; 3) Symbolic or mathematical models.

(4) By Nature of Environment: 1) Deterministic models; 2) Probabilistic models.

(5) By the Time Horizon: 1) Static models; 2) Dynamic models.

Characteristics of a Good Model:

(1) Assumptions should be simple and few.

(2) Variables should be as less as possible.

(3) It should be able to **assimilate** the system environmental changes without change in its framework.

(4) It should be easy to construct.

Constructing the Model:

A mathematical model is a set of **equations** in which the system or problem is described. The equations represent objective function and constraints. Objective function is a mathematical expression of objectives (cost or profit of the operation), while constraints are mathematical expressions of the limitations on the fulfillment of the objectives.

Since model is only an approximation of the real situation, hence it may not include all the variables.

Simplification in Operation Research Models:

While constructing the model, efforts should be made to simplify them, but only up to the extent so that there is no significant loss of accuracy.

Some of the common simplifications are:

(1) **Omitting** certain variables.

(2) Aggregating (or grouping) variables.

(3) Changing the nature of variables e.g., considering variables as constant or continuous.

(4) Changing relationship between variables i.e., considering them as linear or straight line.

(5) Modify constraints.

Operations research can be applied to a variety of use cases, including:

(1) Scheduling and time management.

(2) Urban and agricultural planning.

(3) Enterprise resource planning (ERP) and supply chain management (SCM).

(4) Inventory management.

(5) Network optimization and engineering.

(6) Packet routing optimization.

(7) Risk management.

4.4 Development and Prospect of Operations Research

Without OR, in many cases, we follow these phases in full, but in other cases, we leave important steps out. Judgment and subjective decision-making are not good enough. Thus, industries look to operation research for more objective way to make decisions. It is found that method used should consider the emotional and subjective factors also.

For example, the skill and creative labor are important factors in our business and if management wants to have a new location, the management has to consider the personal feeling of the employees for the location which he chooses.

In its recent years of organized development, O. R. has successfully solved many cases of research for military, the government and industry. The basic problem in most of the developing countries in Asia and Africa is to remove poverty and hunger as quickly as possible. So, there is a great scope for **economist**, **statisticians**, **administrators**, politicians and technicians working in a team to solve this problem by an O. R. approach.

On the other hand, with the explosion of population and consequent shortage of food, every country is facing the problem of optimum **allocation** of land for various crops in accordance with climatic conditions and available facilities. The problem of optimal distribution of water from a resource like a canal for irrigation purposes is faced by developing country. Hence a good amount of scientific work can be done in this direction.

In the field of Industrial Engineering, there is a claim of problems, starting from the **procurement** of material to the dispatch of finished products. Management is always interested in **optimizing** profits.

Hence in order to provide decision on scientific basis, O. R. study team considers various alternative methods and their effects on existing system. The O. R. approach is equally useful for the economists, administrators, planners, irrigation or agricultural experts and statisticians etc.

Operation research approach helps in operation management. Operation management can be defined as the management of systems for providing goods or services and is concerned with the design and operation of systems for the manufacture, transport, supply, or service. The operating systems convert the inputs to the satisfaction of customers need.

Thus, the operation management is concerned with the optimum **utilization** of resources i. e. , effective utilization of resources with minimum loss, underutilization, or waste. In other words, it is concerned with the satisfactory customer service and optimum resource utilization. Inputs for an operating system may be material, machine, and human resource.

O. R. study is complete only when we also consider human factors to the alternatives made available. Operation Research is done by a team of scientists or experts from different related disciplines.

For example, for solving a problem related to the inventory management, O. R. team must include an engineer who knows about stores and material management, a cost accountant a mathematician-cum-statistician. For large and complicated problems, the team must include a mathematician, a statistician, one or two engineers, an economist, computer programmer, psychologist etc.

词汇

生词	音标	释义
simulation	[ˌsɪmjʊˈleɪʃən]	*n.* 模拟；仿真；假装；冒充
probability	[ˌprɒbəˈbɪlɪtɪ]	*n.* 出现的概率；小概率；概率值；相伴概

		率；概率
statistics	[stəˈtɪstɪks]	*n.* 统计数字；统计资料；统计学；（一项）统计数据
components	[kəmˈpəʊnənts]	*n.* 组成部分；成分；部件
game theory		*n.* 博弈论；对策论；博弈理论；游戏理论
optimization theory		*n.* 最优化理论；优化理论
artificial intelligence		*n.* 人工智能；人工智慧；人工智力；人工智能技术
quantitative	[ˈkwɒntɪtətɪv]	*adj.* 数量的；量化的；定量性的
idealised	[aɪˈdɪəlaɪzd]	*adj.* 理想化的
break-even		*n.* 盈亏平衡；盈亏平衡点；收支相抵
assimilate	[əˈsɪmɪleɪt]	*v.* 同化；吸收；融入
equations	[ɪˈkweɪʒənz]	*n.* 方程；等式；方程式；方程组；数学关系式
omitting	[əʊˈmɪtɪŋ]	*v.* 删除；忽略；漏掉；遗漏；不做；未能做

长难句

Operations research (OR) is an analytical method of problem-solving and decision-making that is useful in the management of organizations. In operations research, problems are broken down into basic components and then solved in defined steps by mathematical analysis.

运筹学（OR）是解决问题和决策的一种分析方法，可用于组织的管理。在运筹学中，将问题分解为基本组成部分，然后通过数学分析方法按特定步骤加以解决。

Operation Research model is an idealized representation of the real - life situation and represents one or more aspects of reality.

运筹学模型是现实生活情况的理想化表示，代表了现实的一个或多个方面。

5 Systems Engineering

5.1 The Concept of Systems Engineering

Systems engineering, technique of using knowledge from various branches of engineering and science to introduce technological innovations into the planning and development stages of a system. **Systems engineering is an interdisciplinary field of engineering and engineering management that focuses on how to design, integrate, and manage complex systems over their life cycles.** At its core, systems engineering utilizes systems thinking principles to organize this body of knowledge. The individual outcome of such efforts, an engineered system, can be defined as a combination of components that work in synergy to collectively perform a useful function.

Systems engineering is not so much a branch of engineering as it is a technique for applying knowledge from other branches of engineering and disciplines of science in effective combination to solve a multifaceted engineering problem. It is related to operations research but differs from it in that it is more a planning and design function, frequently involving technical innovation. Probably the most important aspect of systems engineering is its application to the development of new technological possibilities with the specific objective of putting them to use as rapidly as economic and technical considerations permit. In this sense it may be seen as the midwife of technological development.

5.2 Research Objects and Methods of Systems Engineering

(1) Modeling and optimization.

Perhaps the most fundamental technique is the flow diagram, or flowchart, a graphical display composed of boxes representing individual components or subsystems of the complete system, plus arrows from box to box to show how the subsystems interact. Though such a representation is very useful in an initial study, it is, of course, essentially qualitative. A more effective approach in the long run is construction of a so-called mathematical model, which consists of a set of equations, or sometimes simply of tables and curves, describing the interactions within the system in **quantitative** terms. It is not necessary for the mathematical model to be exact, as long as it serves its purpose. It frequently consists of piecewise linear approximations to basically nonlinear situations (i. e., a series of short straight lines that roughly approximate a curve). After the model has been constructed and checked, a number of mathematical techniques can be employed (including straightforward enumeration and computing) to find out what it says about the actual operation of the system. Often these calculations will have a probabilistic or statistical flavor.

When the components or subsystems interact significantly, it may be possible to achieve essentially the same final level of performance in many different ways. Limited performance by one subsystem may be offset by superior performance somewhere else. These optimization studies, called trade-offs, are important in suggesting how to achieve a given result in the most economical manner. They are equally valuable in suggesting whether or not the proposed result is in fact a reasonable goal to aim for. It may be found, for example, that a modest reduction in performance will permit radical savings in overall cost or, conversely, that the postulated equipment is capable of much better performance than is asked of it, at only nominally greater expense. (It may also turn out that the equipment can supply useful functions not originally contemplated. Computing systems, for example, can usually perform extra chores of record keeping at little increased cost.) For all of these reasons, studies of such variables are an important part of systems engineering, both in the early exploratory phases of a project and in the final design.

(2) Identifying objectives.

The **formulation** of suitable objectives for the final system is so important a part of the systems engineering process that it deserves special attention. It is, of course, always possible to state the general objectives of a system in vague or perfectionist terms. A sufficiently clear, precise, and comprehensive statement to serve as a basis for engineering studies, however, is another matter. Unless the situation has been well explored in the past, the real choices are not likely to be obvious when the work begins. Thus, the first task of the systems engineer is to develop as clear a formulation of objectives as possible. This usually involves computations and consultation with others interested in the system. Because the final statement must reflect value judgments as well as purely technical considerations, the systems engineer does not try to do this thinking alone but attempts to serve as a working focus and **catalyst**. Although issues of this sort naturally present themselves with particular force near the beginning of a systems study, they may recur in subsequent steps. The question of objectives is never really out of the systems engineer's mind.

The principal reason why a **satisfactory** statement of objectives may present such a problem is simply that most systems have multiple objectives, often in conflict with one another. In the design of transport aircraft, for example, there are a multitude of desirable characteristics, such as range, speed, payload, and safety, to be maximized, as well as undesirable characteristics, such as noise generation and air pollution, to be minimized. Because the same design cannot do the best job in all of these directions, a compromise achieving the most desirable overall performance is required. The most attractive compromise, which may require both study and ingenuity, is not likely to be found at all until some hard thinking has been done about what characteristics are really needed.

Especially difficult problems in defining objectives may arise when an existing technology is transplanted to some new disciplinary area. An example is the application of electronics such as computer techniques to medicine and education. It seldom happens in such cases that the best system is based on a simple one-for-one substitution, such as direct replacement of a classroom teacher by electronic hardware and computer-assisted instruction materials. It is much more likely that the most effective plan will turn out to be a rather complicated mixture of the old and the new. This conclusion, however, is likely to raise basic issues about the actual objectives of the new system, issues made no simpler by the interdisciplinary nature of the situation.

5.3 The Main Content of Systems Engineering

Systems engineering is to use the system point of view, scientifically and reasonably use cybernetics, information theory, economic management science, modern mathematics, electronic computers and other related engineering technologies to establish a comprehensive optimization system in accordance with the procedures and methods of systems engineering management engineering technology.

The so-called system point of view is to treat the newly studied things as a system; the integrity, purpose, and optimization of the system are the core of system theory. The realization of the control function of the system is based on the feedback theory of cybernetics. This means that a part of the signal should be extracted from the output and fed back to the original end of the system to achieve the purpose of control. The important task of the system is to obtain various information, make judgments, **calculate**, and process, store, transmit, and then output the necessary information. The analysis and evaluation of the system are based on technical and economic indicators. At the same time, the design, development, testing, and application of the system are inseparable from the substance of each branch of management science. The optimization method of modern mathematics mainly means that it uses the theories of many branches of modern mathematics to analyze, summarize and process the various factors that constitute the system, and finally obtain the mathematical model, and the solution of the mathematical model can obtain the optimization The plan is the optimal plan of the system. The use of electronic computers and other related engineering technologies mainly refers to four aspects:

(1) Solving mathematical models.

(2) Processing a large amount of raw data to obtain the information needed for system calculation, operation, management and control.

(3) The computer that composes the link (or sub-system) of the system must accept this secondary information as instructions to complete the management and control functions.

(4) The nature of systems engineering is directly related to relate engineering technology, so it is inevitable to use related engineering technology.

The specific content of systems engineering mainly includes general principles of systems engineering, procedures and methods of systems engineering, system environment research, system model technology, system optimization technology, system analysis technology, system prediction technology, system decision-making technology, system information technology, system control technology, system reliability technology, system simulation technology, ergonomics technology[1].

Figure 5-1 shows single code network diagram.

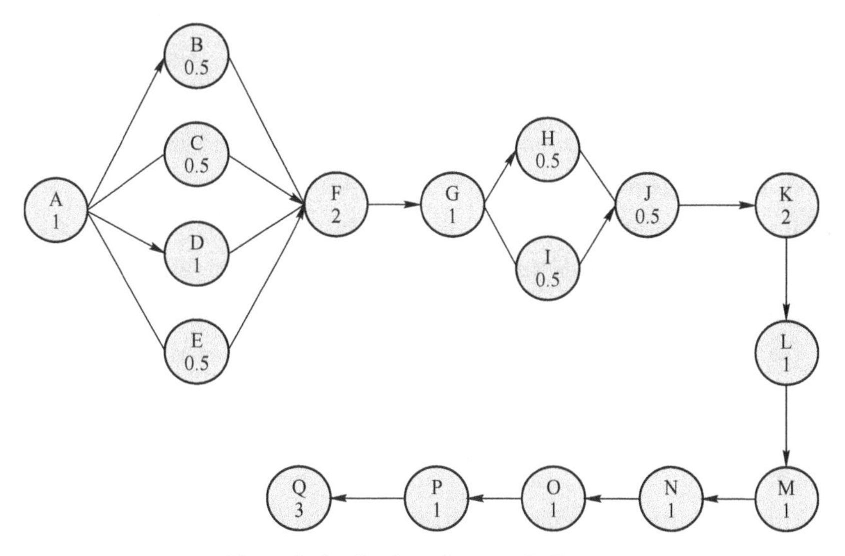

Figure 5-1 Single code network diagram

5.4 Prospects of Systems Engineering

Thus far, systems engineering has been dealt with in relation to two principal fields of application. One field is industry, in which the prospects of a further expansion of systems engineering appear bright. Existing applications furnish many good models, and it seems likely that such extensions can take place without raising many unusual problems. The other major field of application has been in military and space systems, and this may have been the principal force in shaping the systems engineering field. The possibility of new applications of systems ideas in nonmilitary areas of government also has come under consideration in the realm of worldwide basic social and economic problems. On the other hand, systems engineering as practiced in other contexts does not automatically transfer easily to this new environment. General interest in the subject dates, however, only from the late 1960s, and the field is incompletely explored.

In one experiment in the conversion of military systems engineering techniques, a U. S. state government contracted with four large aerospace companies (each of which had a substantial capability in systems engineering) to study the following four topics:

(1) A statewide information-handling system, including a plan for implementation;

(2) A program for the prevention and control of crime and delinquency;

(3) A waste-management problem;

(4) A systems approach to basic transportation problems.

None of the four studies led to proposals that seemed attractive enough to be **implemented** by the state, and, in this respect, the experiment was a disappointment, although in view of the wide scope of the problems attacked and the limited effort called for by the study contracts, the result is not surprising. On the other hand, the experiment was useful in advertising the possibilities of systems analysis as applied to civil problems and in illuminating difficulties that may be encountered in making such applications. The experiment stimulated interest in the civil uses of systems methods both inside and outside the United States.

The potential applications of the systems approach to governmental activities are so numerous and so varied, in both the developed and developing worlds, that an exhaustive catalog would be out of the question. Nevertheless, it may be worthwhile to list a few of the most conspicuous possibilities. The most obvious class is made up of massive engineering attacks on very broad socioeconomic problems. These are the situations that seem to have most in common with the applications of systems methods in developing weapons. They include new transportation systems, comprehensive attacks on pollution, and radical reconstruction of urbanareas. A concrete issue is the problem of power-plant location, an urgent question in many advanced and developing countries. The systems overtones are obvious. Generating stations are customarily interconnected so that a new plant has an impact on the availability of power over a considerable region, and, of course, the effects of thermal and atmospheric pollution from a given plant may also be widespread.

Other applications of systems analysis in the social sphere tend generally to be smaller and more easily treated. One class consists of the extension of military budgeting and methods of financial control to the nonmilitary world. Another application has been the use of systems analysis to support the technical aspect of foreign-aid programs. Other fields include the possible application of specific items of new technology in such areas as crime detection, firefighting, and traffic control. Still other studies involve specific aspects of such subjects as housing and other types of building construction. Such studies attempt to be useful rather than broad or necessarily definitive for all time to come.

The applications of systems analysis in civil government obviously still have far to go before their potentialities are exhausted. On the other hand, there are many reasons why these potentialities can be realized only slowly, if at all. Some of them are related to the inherent difficulty of the problems presented—the wide range of both technical and social considerations that may enter certain decisions, for example. Others reflect some of the common characteristics of governmental structure, the necessary **bureaucratization** of functions, for example, or the frequent problem of overlapping jurisdiction. Still other problems reflect the fact that existing systems analysts are trained preponderantly in the physical sciences and engineering and thus may not be well matched to the socioeconomic issues they are likely to confront, though most systems analysis groups working in socioeconomic questions try to balance their strength by adding appropriate missing skills. The most common problem, however, is probably simply the need to build up an adequate basis for mutual cooperation between systems analysts and government.

As such an evolution proceeds, there may be an increasing tendency for individual systems analysts to become identified with the **substantive** area in which they work and to lose their special relations to systems analysis as a distinctive field. **Thus, it may be the ultimate fate of systems analysis to disappear as a separate field and instead become an important constituent of the planning function required in many parts of modern society[2].**

词汇

生词	音标	释义
interdisciplinary	[ˌɪntədɪsəˈplɪnərɪ]	*adj.* 多学科的；跨学科的
formulation	[ˌfɔːmjʊˈleɪʃn]	*n.* （药品、化妆品等的）配方；配方产品；构想；（想法的）阐述方式，表达方法
satisfactory	[ˌsætɪsˈfæktərɪ]	*adj.* 令人满意的；够好的；可以的
calculate	[ˈkælkjʊleɪt]	*v.* 计算；核算；预测；推测
implemented	[ˈɪmplɪmentɪd]	*v.* 使生效；贯彻；执行；实施
socioeconomic	[ˌsəʊsɪəʊˌɛkəˈnɒmɪk]	adj. 社会经济的
bureaucratization	[bjʊərəkrətɑɪˈzeɪʃn]	*n.* 官僚主义化
substantive	[səbˈstæntɪv]	*adj.* 实质性的；本质上的；重大的；严肃认真的

长难句

Systems engineering is an interdisciplinary field of engineering and engineering management that focuses on how to design, integrate, and manage complex systems over their life cycles.

系统工程是工程和工程管理的一个跨学科领域，其重点是如何在整个生命周期中设计、集成和管理复杂的系统。

Systems engineering is to use the system point of view, scientifically and reasonably use cybernetics, information theory, economic management science, modern mathematics, electronic computers, and other related engineering technologies to establish a comprehensive optimization system in accordance with the procedures and methods of systems engineering Management engineering technology.

系统工程就是要从系统的角度出发，科学地、合理地运用控制论、信息论、经济管理科学、现代数学、电子计算机等相关工程技术，按照系统工程的程序和方法，建立一套综合的优化系统的管理工程技术。

Thus far, systems engineering has been dealt with in relation to two principal fields of application. One field is industry, in which the prospects of a further expansion of systems engineering appear bright. Existing applications furnish many good models, and it seems likely that such extensions can take place without raising many unusual problems. The other major field of application has been in military and space systems, and this may have been the principal force in shaping the systems engineering field.

迄今为止，系统工程已经在两个主要应用领域进行了实践。一个领域是工业，其中系统工程进一步扩展的前景似乎十分光明。现有的应用程序构建了许多好的模型，并且可以进行此类扩展而不会引起许多异常的问题。另一个主要的应用领域是军事和太空系统，这可能是系统工程应用领域的主要方向。

Reference

[1] Emmanuel C, Bertrand R, Virginie G. Research methodology for systems engineering: some recommendations [J]. IFAC-PapersOnLine, 2016, 49: 1567-1572.

[2] Emmanuel C, Bertrand R, Virginie G. Research methodology for systems engineering: some recommendations [J]. IFAC-PapersOnLine, 2016, 49: 1578-1580.

6 Work Study

6.1 Overview of the Work Study

6.1.1 Introduction of Work Study

According to ILO — International Labor Organization — work study is "**a term used to embrace the techniques of method study and work measurement which are employed to ensure the best possible use of human and material resources in carrying out a specified activity.**" In other words, "work study is a tool or technique of management involving the analytical study of a job or operation." Work study helps to increase productivity.

6.1.2 Objectives of Work Study

(1) Work study brings higher productivity;
(2) Work study improves existing method of work for which cost becomes lower;
(3) It eliminates wasteful elements;
(4) It sets standard of performance;
(5) It helps to use plant and human more effectively;
(6) It improves by saving in time and loss of material also.

In nutshell work study is mainly concerned with the examination of human work. In fact, planning is not possible unless one knows how long it will take to do a particular job. Thus, time is very important to the manufacturer who must keep promising, to estimate quantities and to other industrial and business arrangements or organizations.

Work study is not a theoretical concept but essentially a practical one and deals with human beings who have their own attitude and style of working. So, the success of work study is dependent upon the relations between the labor/employees and the management.

Work study involves lot of changes in various working methods. Since the manpower in general does not like changes but prefers to continue as already doing, so there will always be a tendency to resist any modification or new method suggested by work study people (officers/workers) and the manpower and the workers have confidence in the ability, integrity and fair-mindedness of work study man, there is a good chance that sound proposals will be accepted willingly by the manpower.

6.1.3 Definition and Concept of Work Study

Generally, work study is used to describe a complete set of techniques with the help of which work can be simplified, standardized, and measured.

When it is possible to simplify the existing work or new methods are designed and introduced such that the task/activity becomes simpler than following advantages are possible:

(1) More production with less effort so goods/products are available at cheaper rates;

(2) Better equipment utilization shall lead to marked increase in the total production without addition of new resources, thus productivity may improve.

These advantages are possible through the willing cooperation of the people engaged in production work. In view of these far-reaching benefits, work study has become an important tool of management.

In industries work study is considered as a tool of improving productivity by way of:

(1) Resource utilization to a satisfactory level;

(2) Capital investment to introduce latest technology;

(3) Better management of the system.

British Standard Institution defines work study as a generic term for those techniques particularly "Method study" and "Work Measurement" which are used in the examination of work in all its contexts, and which leads systematically to the investigation of all the factors which affect the efficiency and economy of the situation being reviewed in order to incorporate improvements at various levels.

Work study may be defined as "The systematic critical, objective and imaginative examination of all factors governing the operational efficiency of any specific activity in order to achieve/effect improvement."

Thus, work study is the investigation by means of a consistent system of the work done in an organization in order to achieve the best possible utilization of resources i. e., man, machines, and materials available. Every organization tries to achieve best quality production of various products in the minimum possible time.

6.1.4 Need for Work Study

Principles of work study have been used since long to identify the improvements to be incorporated, when industrial set up was simple and involved lesser problems. The industries of today with increased complexities and modernization naturally demand a more systematic approach like work study in its present form for solution of various problems.

Work study is most valuable tool of management because:

(1) It is a direct means of improving productivity of the system involving very less or no cost.

(2) The approach is simple, systematic, consistent, and based on handling of facts. Thus, the part played by opinions in taking decisions is minimized.

(3) No factor affecting the efficiency of operation is overlooked in this approach.

(4) It provides most accurate means of setting standards of performance which are helpful in the process of production planning and control.

(5) Application of work study result in immediate savings.

(6) It is a universal tool for management.

(7) It is a most penetrating tool of investigation available to the management of the industrial unit.

6.1.5 Analytical Techniques for Work Study

The analytical techniques commonly used in Work Study are: "5W1H" "ECRS".

5W1H is the abbreviation summarizing the following six questions: What? Who? Where? When? Why? How? This method consists of asking a systematic set of questions to collect all the data necessary to draw up a report of the existing situation with the aim of identifying the true nature of the problem and describing the context precisely. "**5W1H**" is shown in Figure 6-1.

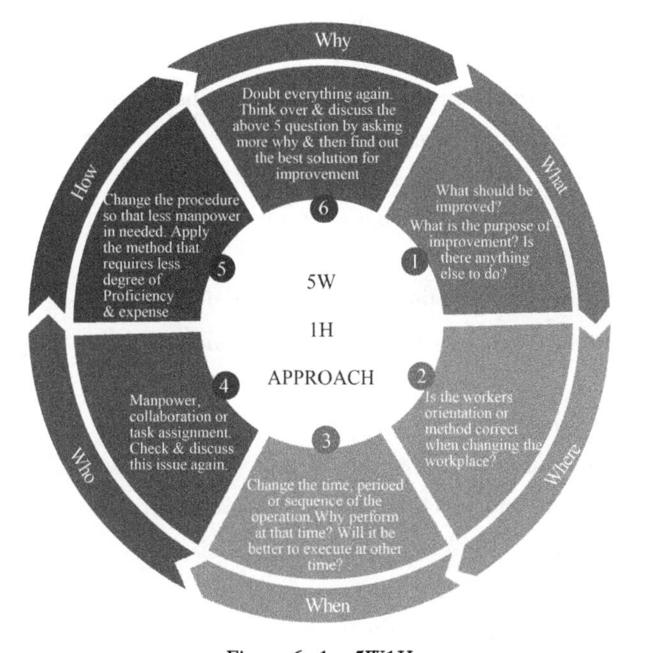

扫一扫查看彩图

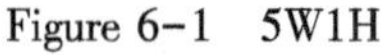
Figure 6-1 5W1H

"ECRS" stands for Eliminate, Combine, Rearrange, and Simplify. This Lean technique is primarily used to reduce or remove wasteful steps entirely from anything regarding manufacturing processes or even office procedures. It might even be helpful to put together a value stream map as well to identify necessary and complex steps within production lines. When implementing ECRS, complex and time-consuming tasks are reviewed with the goal of successfully applying any one or all of the strategies in the ECRS acronym. The end result aims to streamline any process for workers and create a more efficient business as a whole. "**ECRS**" is shown in Figure 6-2.

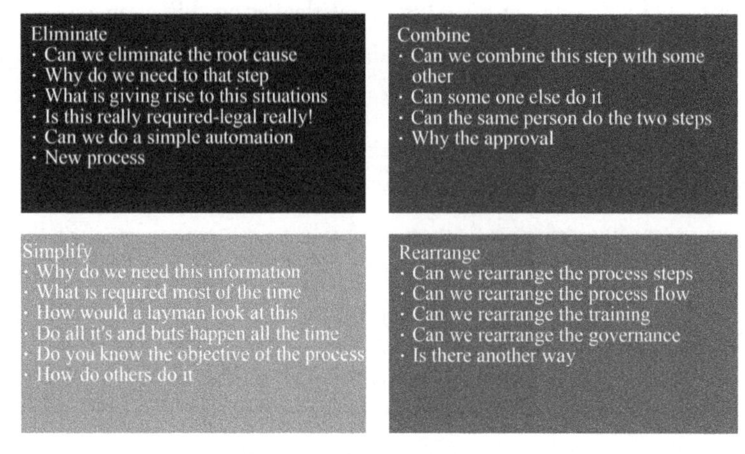

扫一扫查看彩图

Figure 6-2 ECRS

6.1.6 Procedure/Steps Involved in Work Study

The basic procedure of work study is as follows:

(1) Select the job or process to be suited.

(2) Record from direct observation everything that happens in order to obtain data for analysis.

(3) Examine the recorded facts critically and challenge everything that is done, considering in turn: the purpose of activity, the place where it is.

(4) Performed, the sequence in which the elements are performed, the person who is doing it, the means by which it is done.

(5) Develop the most economic methods, considering all the circumstances.

(6) Measure the amount of work involved in the method used and calculate a "standard time" for doing it.

(7) Define the new method and the related time.

(8) Install the new method and time as agreed standard practices.

(9) Maintain the new standard practice by proper control procedures.

6.1.7 Advantages of Work Study

(1) It is direct means of improving productivity.

(2) It results in uniform and improved production flow.

(3) It reduces the manufacturing cost.

(4) With its help fast and accurate delivery dates are possible.

(5) It provides better service and consumer satisfaction.

(6) It improves employee-employer relations.

(7) It provides job satisfaction and job security to workers.

(8) Better working conditions are possible for workers.

(9) It is most important tool of analysis and can help in providing better wages to workers on scientific basis.

(10) Most accurate method and yet provides a sound basis for production planning, control and incentives for manpower.

(11) Everyone concerned with industries is benefited from it such as work.

6.1.8 Techniques of Work Study

Basically, there are two techniques:

Method study and work measurement are shown in Figure 6-3.

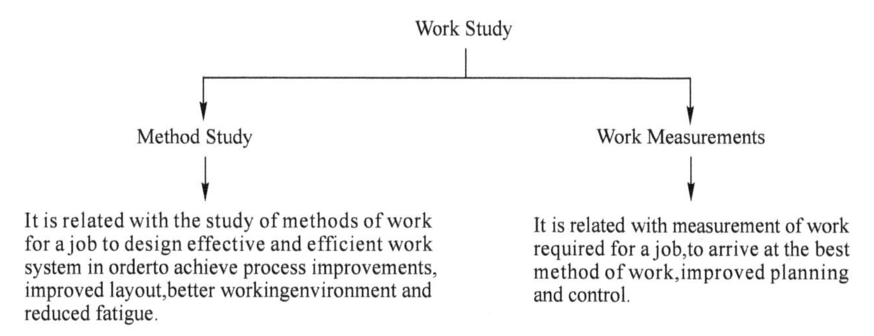

Figure 6-3 Work Study

Thus, work study is the term used to embrace the techniques of Method Study and Work Measurement which are used to ensure the best utilization of manpower and material resources in carrying out specified activity.

The **sequential** order of the correct procedure to be adopted for having effective or purpose-oriented results of method study include the following:

(1) Select the work/procedure to be analyzed.

(2) Record all the relevant information related with the existing work system with the help of various recording devices or techniques.

(3) Make critical examination of collected data/facts.

(4) Develop and improve the method which is economical and practical after giving due consideration to the alternative method possible.

(5) Install the new selected method with proper instructions.

(6) Maintain the latest standards of methods through periodic verification etc.

6.2 Method Study

6.2.1 Concept and Definition

Method study is basically conducted to simplify the work or working methods and must go towards higher productivity. It is always desirable to perform the requisite function with desired goal minimum consumption of resources. Method signifies how a work is to be done i.e., description of how we consume resources in order to achieve our target?

Methods are integral part of work accomplishment and signify:

(1) How well our methods utilize the limited available resources such as manpower, machines, materials, and money.

(2) How our methods physically affect production output of the unit.

(3) The quality of output obtained by application of our methods.

Thus, methods can determine the amount of input materials, time power and money consumed. So, methods may be considered the core where one can attempt to reduce the consumption of resources thereby reducing cost per unit output through utilization of proper methods. The method design can decide the cost and quality of output produced.

Method Study may be defined as: **"A procedure for examining the various activities associated with the problem which ensures a systematic, objective and critical evaluation of the existing factors and in addition and imaginative approach while developing improvements"**.

There are three aspects of its application:

(1) Method study proper is concerned with broad investigation and improvement of a shop/section, the layout of equipment and machines and the movement of men and materials.

(2) Motion study is a more detailed investigation of the individual worker/operator, layout of his machines, tools, jigs and fixtures and movement of his limbs when he performs his job. The ergomics aspect i. e., study of environment, body postures, noise level and surroundings temperature also form part of investigation.

(3) Micro motion study i. e., much more detailed investigation of very rapid movements of the various limbs of the worker.

So, motion study is an analysis of the flow and processing of material and the movements of men through or at various workstations. Thus, motion study analyses the human activities which make up an operation. Whereas method study or methods analysis has been defined as: "systematic procedure for the critical analysis of movements made by men, materials and machines in performing any work".

Now because by definition method study includes the study of all facets of human work and all factors affecting the work so motion study be considered as a part of method study.

6.2.2 Scope of Method Study

The task of work simplification and compatible work system design concerns the followings:

(1) Layout of shop floor and working areas or workstations;

(2) Working conditions i. e., ergomics etc;

(3) Handling distances (material movement);

(4) Tooling and equipment used;

(5) Quality standards to be achieved;

(6) Operators and operations in achieving the production targets;

(7) Materials to be used;

(8) Power required and available;

(9) Work cycle time;

(10) Working processes.

All these factors are related to method study and possible improvements may be:

(1) Short term:

The improvements which can be introduced quickly and economically. These may be concerned with management and work force.

(2) Long term:

The improvements which are not acceptable to management at present and which require good investment. Improvement approach to method design is essential since a method describes how resource are to be used in order to convert them into desired output (final products) in order to accomplish the purpose through a network of facilities.

Operation and route sheets of production process contains in instructions that how a particular product/component can be manufactured. This usually contains the details about time required to perform the required operation.

The target is the minimization of production costs by affecting the consumer's acceptability by incorporating changes or by developing requisite resigns. But the design engineer will not be responsible for actual implementation of method designed by him. Likewise, the Process Engineer will try to select best methods which have most economical sequence of operations by using most efficient infrastructure facilities (may be machines) with processing minimum time.

6.2.3 Steps of Method Study

(1) Select (the work to be studied);

(2) Record (all relevant information about that work);

(3) Examine (the recorded information);

(4) Develop (an improved way of doing things);

(5) Install (the new method as standard practice);

(6) Maintain (the new standard proactive).

6.2.4 The Content of Method Study

(1) Process analysis;

(2) Operation analysis;

(3) Motion analysis.

6.2.5 Process Analysis

6.2.5.1 Definition

Process analysis can be understood as the rational breakdown of the production process into different phases, that turns input into output. It refers to the full-fledged analysis of the business process, which incorporates a series of logically linked routine activities, that uses the resources of the organization, to transform an object, with the aim of achieving and maintaining the process excellence.

Process analysis (as shown in Figure 6-4) is nothing but a review of the entire process flow of an organization to arrive at a thorough understanding of the process. Further, it is also helpful to set up targets for the purpose of process improvement, which is possible by eliminating unnecessary activities, reduce wastage and increasing efficiency. Thus, it ultimately ends up improving the overall performance of the business activities.

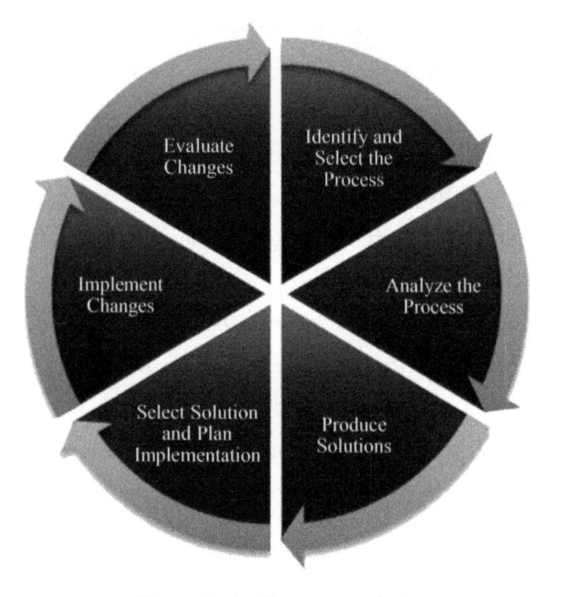

扫一扫查看彩图

Figure 6-4 Process Analysis

6.2.5.2 Objectives of Process Analysis

(1) Identify the factors that make it difficult to understand the process;

(2) Ascertain completeness of the process;

(3) Remove bottlenecks;

(4) Find redundancies;

(5) Ascertain the allocation of resources;

(6) Check out process time.

Understanding, Quality and Efficiency are the three basic criterion, through which one can analyze the process and determine the areas that require change.

6.2.5.3 Steps Involved in Process Analysis

(1) Interview major participants of the process: Discuss the participants about what they do, why they do and how they do it. Identify the information and inputs required by the workers to perform the task assigned to them. Research about the source of input and outputs of each task.

(2) Carry out group discussion: Group interview and brainstorming session are conducted, with

the aim of generating ideas, validating, and refining the information collected, at the first step.

(3) Identify bottlenecks and redundancies: Find out the bottlenecks in each task that causes delay and various measures to remove it. Further, identify the unnecessary activities, whose elimination can ease the process.

(4) Create **Sketch**: Make a sketch right from the scratch of the entire process, depending upon the business process requirements, which came into light after interviews and discussions.

(5) Compare: At the end, compare the latest process flow with the previous one, and mark the areas that require changes, as per the research conducted.

Process Analysis is a methodical approach to enhance the understanding and redesigning of the workflow of the organization. It acts as a tool to maintain and improve the business processes and also help in attaining the incremental to transformational benefits, such as cost reduction, optimum utilization of resources, effective human resource allocation and process efficiency.

6.2.6 Operation Analysis

6.2.6.1 Definition

An operation analysis is a procedure used to determine the efficiency of various aspects of a business operation. Most reports include a careful scrutiny of a company's production methods, material costs, equipment **implementation** and workplace conditions. Professional consultants are often brought in from outside a company to perform an unbiased operational analysis, which provides a company with hard data concerning waste issues and operational risks. Many companies use the information from such an analysis to decide on what changes need to be made to improve operations.

6.2.6.2 Basic Requirements

(1) Through deletion, merging, and simplification, the total number of operations is minimized, the process sequence is optimal, and each operation is simple and easy.

(2) Make reasonable use of muscle groups to prevent certain muscle groups from straining due to too frequent movements. Give full play to the role of both hands, balance the load of both hands, avoid holding objects with your hands for a long time, and try to use tools.

(3) The machine is required to complete more work, such as: automatic feed, retract, stop, automatic detection, automatic tool change, etc. For mass production, an automatic feeding and unloading device should be designed. Improve parts box or parts placement method.

(4) Reduce operating cycles and frequency. Reduce the number of transportation and transfer of materials, shorten the transportation and movement distance, and make transportation and movement convenient and easy.

(5) Improve equipment, tooling and workstation equipment, material specifications or technology, and adopt economical cutting consumption.

(6) The workplace should have enough vacancy, so that the operator has enough room for maneuver.

(7) Eliminate unreasonable idle time, try to realize the synchronization of man-machine work, and make certain preparations, work placement work, and auxiliary work carried out in flexible hours.

In summary, through job analysis, the goal of rationalizing the structure of the job, reducing the labor intensity of the operator, reducing the time consumption of the job, ensuring the quality of production, and improving the efficiency of the job should be achieved.

6.2.6.3 Features

(1) Analyze and improve the work on a job site in detail as the object and goal of analysis.

(2) The basic method of application analysis is to analyze the human-oriented operating system.

(3) The content of the analysis is various factors that affect the efficiency and quality of the operation. It usually includes working methods, raw materials, equipment and tools, working environment conditions, etc.

6.2.6.4 Types

The types of operation analysis can be divided into:

(1) Man-machine operation analysis: In the machine work process, investigate and understand the relationship between machine operation and worker operation during the operation cycle, so as to make full use of the energy and balance operation of the machine and worker.

(2) Joint job analysis: Record and analyze the various actions and their relationships of each object in a work program and achieve the purpose of shortening the cycle by rationally deploying the work of each work object, canceling idle or waiting time.

(3) Two-handed operation analysis: investigate and understand how to use both hands for actual operation and know how the operator uses both hands effectively.

6.2.7 Motion Analysis

6.2.7.1 Definition

Motion analysis is a measuring technique used in computer vision, image processing and high-speed photography applications to detect movement. The objectives of motion analysis are to detect motion within an image, track an object's motion over time, group objects that move together and identify the direction of motion. Specific techniques for implementing motion or movement analysis include electromyography, background segmentation and differential equation models.

Motion analysis software can help organizations implement aspects of motion analysis with minimal effort. Once connected to motion capture devices, the software can visually display object tracking information for further examination and manipulation.

6.2.7.2 How Motion Analysis Works

The basic function of motion analysis is to compare two or more consecutive images captured by sensors or cameras to return information on the apparent motion in the images. This is usually done by programming the recording device to produce binary images based on movement. All of the image points, or pixels, that correspond with motion are set to a value of 1 while stationary pixels are set to 0. The resulting image can be processed even further to remove noise, label objects and group neighboring values of 1 into a singular object.

The data produced by motion analysis tools often correlates to a specific image at a specific point in time based on its position in the sequence. Therefore, the motion capture data is time-dependent, which is a crucial component in most tracking applications.

6.2.7.3 Uses of Motion Analysis

Motion analysis is used in a variety of fields and applications, including:

(1) Manufacturing—Motion analysis can be applied to the manufacturing process through software that monitors and analyzes supply chains for inefficiencies or malfunctions. Similarly, motion analysis can be used by manufacturers to conduct product safety, collision, or efficiency tests.

(2) Video surveillance—Human activity recognition through motion analysis is commonly used for security monitoring and surveillance purposes.

(3) Healthcare and physical therapy—Motion analysis can help healthcare providers track muscle activity, perform gait analysis, and diagnose potential mobility issues. Treatment for patients with cerebral palsy, spine bifida, muscular dystrophy and joint issues can involve regular motion analysis tests in a laboratory.

(4) Autonomous vehicles—Motion analysis systems can be used in self-driving cars to aid with traffic navigation and obstruction identification.

(5) Biological sciences—Specific motion analysis software can be used to count and track tiny particles such as bacteria and viruses.

6.2.7.4 The Method of Motion Analysis

(1) Therbligs;

(2) Imagery analysis.

6.2.7.5 Therbligs

Therbligs are 18 kinds of elemental motions that make up a set of fundamental motions required for a worker to perform a manual operation or task. They are used in the study of motion economy in the workplace. A workplace task is analyzed by recording each of the therblig units for a process, with the results used for optimization of manual labor by eliminating unneeded movements. **Therbligs** are shown in Figure 6-5.

Figure 6-5 Therbligs

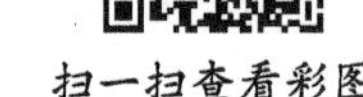

(1) Transport empty unloaded: receiving an object with an empty hand. (Now called "Reach")

(2) Grasp (G): grasping an object with the active hand.

(3) Transport loaded (TL): moving an object using a hand motion.

(4) Hold (H): holding an object.

(5) Release load (RL): releasing control of an object.

(6) Preposition (PP): positioning and/or orienting an object for the next operation and relative to an approximation location.

(7) Position (P): positioning and/or orienting an object in the defined location.

(8) Use (U): manipulating a tool in the intended way during the course working.

(9) Assemble (A): joining two parts together.

(10) Disassemble (DA): separating multiple components that were joined.

(11) Search (Sh): attempting to find an object using the eyes and hands.

(12) Select (St): choosing among several objects in a group.

(13) Plan (Pn): deciding on a course of action.

(14) Inspect (I): determining the quality or the characteristics of an object using the eyes and/or other senses.

(15) Unavoidable delay (UD): waiting due to factors beyond the worker's control and included in the work cycle.

(16) Avoidable delay (AD): waiting within the worker's control which causes idleness that is not included in the regular work cycle.

(17) Rest (R): resting to overcome a fatigue, consisting of a pause in the motions of the hands and/or body during the work cycles or between them.

(18) Find (F): A momentary mental reaction at the end of the search cycle. Seldom used.

6.2.7.6 Imagery Analysis

(1) Definition of image analysis:

Image analysis refers to a technology that uses a camera, etc. to take pictures of operational actions, record human movements, and perform analysis and research through projections. According to different shooting speeds, audio-visual analysis is divided into subtle motion analysis and slow-motion analysis. It has the characteristics of high accuracy, repeatability, and easy operation. It is suitable for operation analysis with short product cycle, high complexity, and complex actions.

(2) Uses of image analysis:

Make up for the limitation of human analysis ability;

Determine the time value of the operation cycle that is difficult to observe (such as the operation with less repetitiveness and long cycle, etc.);

Used to record the true status of the scene;

Present the situation of the job site in other places for everyone to discuss;

Persuade and explain to relevant personnel.

(3) Image analysis method:

There are two methods of image analysis, namely slow-motion analysis and subtle motion image analysis.

6.2.7.7 Principles of Motion Economy

The principles of motion economy form a set of rules and suggestions to improve the manual work in manufacturing and reduce fatigue and unnecessary movements by the worker, which can lead to the reduction in the work-related trauma. The principles of motion economy are grouped under four headings:

(1) Use of the human body:

When possible:

The two hands should begin and complete their movements at the same time.

The two hands should not be idle at the same time except during periods of rest.

Motions of the arms should be symmetrical and in opposite directions and should be made simultaneously.

Hand and body motions should be made at the lowest classification at which it is possible to do the work satisfactorily.

Momentum should be employed to help the worker but should be reduced to a minimum whenever it has to be overcome by muscular effort.

Continuous curved movements are to be preferred to straight-line motions involving sudden and sharp changes in direction.

"Ballistic" (i.e., free-swinging) movements are faster, easier, and more accurate than restricted or controlled movements.

Rhythm is essential to the smooth and automatic performance of a repetitive operation. The work should be arranged to permit easy and natural rhythm whenever possible.

Work should be arranged so that eye movements are confined to a comfortable area without the need for frequent changes of focus.

(2) Arrangements of the workplace:

Definite and fixed stations should be provided for all tools and materials to permit habit formation.

Tools and materials should be pre-positioned to reduce searching.

Gravity fed bins and containers should be used to deliver the materials as close to the point of use as possible.

Tools, materials, and controls should be located within the maximum working area and as near to the worker as possible.

Materials and tools should be arranged to permit the best sequence of motions.

"Drop deliveries" or ejectors should be used wherever possible so that the operator does not have to use his hands to dispose of the finished work.

Provision should be made for adequate lighting and a chair of the type and height to permit good posture should be provided.

The height of the workplace and seat should be arranged to allow alternate standing and sitting.

The color of the workplace should contrast with that of the work and thus reduce eye fatigue.

(3) Design of tools and equipment:

The hands should be relieved from 'holding' the work piece where this can be done by a jig, fixture, or foot-operated device.

Two or more tools should be combined wherever possible.

Where each finger performs some specific movement, as in typewriting, the load should be distributed in accordance with the inherent capacities of the fingers.

Handles such as those on cranks and large screwdrivers should be designed so as to permit as much of the surface of the hand as possible to come into contact with the handle. This is especially necessary when considerable force has to be used on the handle.

Levers, crossbars, and hand-wheels should be placed so that the operator can use them with the least change in body position and the greatest mechanical advantage.

(4) Time conservation:

Even a temporary delay of work by a man or machine should not be encouraged.

Machine should not run idle; it is not desirable that a lathe machine is running, and its job is rotating but no cut is being taken.

Two or more jobs should be worked upon at the same time or two or more operations should be carried out on a job simultaneously if possible.

Number of motions involved in completing a job should be minimized.

6. 3 Work Measurement

6. 3. 1 Meaning and Definition of Work Measurement

Work measurement is concerned with the determination of the amount of time required to perform a unit of work. Work measurement is very important for promoting productivity of an organization. It enables management to compare alternate methods and also to do initial staffing. Work measurement provides basis for proper planning.

Since it is concerned with the measurement of time it is also called "Time Study". The exact examination of time is very essential for correct pricing. To find the correct manufacturing time for a product, time study is performed. To give competitive quotations, estimation of accurate labor cost is very essential. It becomes a basis for wage and salary administration and devising incentive schemes.

Work measurement has been defined by British Standard Institution as, "The application of techniques designed to establish the time for a qualified worker to carry out a specified job at a defined level of performance". This time is called standard or allowed time. Time study may also be defined as "the art of observing and recording the time required to do each detailed element of an industrial operation".

6. 3. 2 Objectives of Work Measurement

(1) To compare the times of performance by alternative methods.

(2) To enable realistic schedule of work to be prepared.

(3) To arrive at a realistic and fair incentive scheme.

(4) To analyze the activities for doing a job with the view to reduce or eliminate unnecessary jobs.

(5) To minimize the human effort.

(6) To assist in the organization of labor by daily comparing the actual time with that of target time.

6. 3. 3 Uses of Work Measurement

(1) Work measurement is used in planning work and in drawing out schedules.

(2) Work measurement is used to determine standard costs.

(3) Work measurement is used as an aid in preparing budgets.

(4) It is used in balancing production lines for new products.

(5) Work measurement is used in determining machine effectiveness.

(6) To determine time standards to be used as a basis for labor cost control.

(7) To establish supervisory objectives and to provide a basis for measuring supervisory efficiency.

(8) To determine time standards to be used for providing a basis for wage incentive plans.

6.3.4 Techniques of Work Measurement

Work measurement is investigating and eliminating ineffective time. It not only reveals the existence of ineffective time. But it can be used to set standard times for carrying out the work so that ineffective time does not evolve later. It will be immediately found out by the increased standard time. For the purpose of work measurement, work may be regarded as repetitive work and non-repetitive work.

The principal techniques of work measurement are classified under the following heads:

(1) Time study;

(2) Work sampling;

(3) Pre-determined motion time system;

(4) Analytical estimating.

6.3.5 The Content of Work Measurement

(1) Stopwatch time study;

(2) Work sampling;

(3) Predetermined Motion Time Systems (PMTS);

(4) Standard data.

6.3.6 Stopwatch Time Study

6.3.6.1 Meaning of Stopwatch Time Study

Stopwatch Time Study is one of the equipment used for Time Study. It is employed for measuring the time taken by an operator to complete the work. Stopwatch used for time study purpose should be very accurate and preferably be graduated in decimals so that it can recover even up to 0.01 minute.

A large hand in the stopwatch is revolved at a speed of one revolution per minute. The dial of the stopwatch is divided into 100 equal divisions. The small hand inside the stopwatch revolves at a speed of one revolution in 30 minutes.

6.3.6.2 Procedure of Stopwatch Time Study

The stopwatch procedures for collecting Time Study Data are listed below:

(1) Analyze the job to establish the quality to be achieved in the job.

(2) Identify key operations to be timed in the job.

(3) Get improved procedure from the method study department.

(4) Organize resources and explain the objectives of time study to the worker and supervisor.

(5) Explain details to worker about improved working procedure.

(6) Break operation into elements to separate the constant elements from variable elements.

(7) Observe and record the time taken by an operator.

(8) Determine for number cycles to be timed and then the average time or representative time can be found out.

(9) Rate of performance of the worker during observation.

(10) Calculate normal time from observed time by using performance rating factor.

(11) Add process allowance rest and personal allowance and special allowances to the normal time in order to obtain standard time or allowed time.

(12) Standard time determination by adding normal time and allowances.

6.3.6.3 Equipment's of Stopwatch Time Study

Stopwatch is one of the important timing devices used for measuring the time taken by a worker to complete a job. Stopwatch is an accurate time measuring equipment which can normally run continuously for half an hour or one hour and record the time by its small hand. One revolution of the big hand records one minute. Even the scale covering one minute may be calibrated into intervals of 1/100th of a minute.

There are three types of stop watches and they are:

(1) Non-fly back stopwatch.

(2) Fly back stopwatch.

(3) Split hand or split-second type stopwatch.

6.3.6.4 Basic Procedures for Stopwatch Time Study

(1) Receive the request-for time study.

(2) Obtain the cooperation of the departmental foreman or supervisor.

(3) Select an operator and obtain his cooperation.

(4) Determine whether the job or operation is ready for time study.

(5) Obtain and record all necessary information.

(6) Divide the operation into elements and record complete description of the method.

(7) Observe and record element times.

(8) Present the data.

6.3.7 Work Sampling

Work-sampling is defined as a technique for determining and predicting the total or the proportion of the time consumed by a specified activity. It is dependent upon the observations that have been made over a while to record the frequency of the events that are being performed and the happenings in that instant.

6.3.7.1 Meaning of Work Sampling

Work-sampling is described as a tool used by employers to determine the time which an employee spends on a specific activity or task.

Work-sampling is a statistical concept which permits recognition, analysis and enhancement of job responsibilities, organizational workflows, performance competencies and tasks.

An essential use of the work-sampling method is predicting the standard time for a manufacturing task conducted manually. It is used in several processes like telemarketing, manufacturing, and customer service.

6.3.7.2 Features

The general characteristics of work-sampling method are as follows:

(1) It requires sufficient time to complete the study.

(2) It is feasible for the work-sampling method to study multiple workers at a time instead of one single worker or a small group.

(3) The cycle time of work-sampling is generally lengthy.

(4) The work cycle is non-repetitive.

6.3.7.3 Applications

The applications of work-sampling process are as follows:

(1) Work-sampling enables a fair share of job distribution amongst the workforces.

(2) It is applied to find an estimate about delay times that are unavoidable for deciding about the delay allowances.

(3) The data that is available from the work-sampling method has proved a great help in the process of production planning.

(4) A work-sampling method is a helpful tool for the administration as it helps in evaluating the efficiency levels of the various departments in the organization.

(5) The work-sampling assist the management in an organization to find all the data and information of idle time and its cause.

(6) Work sampling can be applied for the utilization of cranes, machine tools, trucks, etc.

(7) It is used to get an estimate of the percentage of time consumed by various job activities like inspection, repair, and supervision.

(8) Work sampling is a useful tool that is applicable in finding time standards for repair work, office work, maintenance work, etc.

6.3.7.4 Advantages

The advantages of work-sampling are as follows:

(1) It is possible to interrupt the study related to work-sampling at any given time without any impact on the results of the study.

(2) It is easy to identify and eliminate uneconomical activities through the process of work-sampling.

(3) An organization need not hire any experts to conduct the process of work-sampling as it can be carried out by anyone, even an employee with limited training and knowledge.

(4) A single analyst is enough for group operations.

(5) One of the advantages of work-sampling is that it provides the unbiased result as the workmen are not subjected to close observation.

(6) Work-sampling is less time consuming as well as economical by nature because it is possible to study more than one workforce at the same time.

Moreover, the duration of time taken for the study is only a few hours and this is also the reason why it is less time consuming.

(7) Another benefit of work-sampling is that the observer need not be present himself for the whole process as it automatically records its findings.

(8) It is possible to study teamwork by work-sampling and not by the time study.

(9) An essential advantage of work-sampling is that it helps in reducing the clerical time.

(10) Numerous activities that are costly and cannot be measured by time study can be easily measured by work-sampling.

(11) It is less tedious and does not cause fatigue.

(12) The work sampling process does not need a timing device or a stopwatch.

6.3.7.5 Disadvantages

The disadvantages of work sampling method are as follows:

(1) Work sampling does not have any provision for small delays or breakdown of activities.

(2) If an organization is interested in work sampling of one employee or small groups of a worker, then the process will not prove economical. It is also not cost-effective for short-cycle jobs.

(3) One of the disadvantages of the work sampling method is that the results are often inaccurate if the observations are insufficient or limited in nature.

(4) In some cases, the results from work sampling are not accurate because the working men who are being studied may change their usual working method on seeing an observer.

(5) One of the limitations of work sampling method is that it does not record the speed of working of a worker.

(6) Workers and management can't understand the concept of work sampling as effortlessly as time study.

6.3.7.6 Errors

Several errors may occur during the work sampling process, and it is essential to avoid them. Some common ones are as follows:

(1) Bias in working samples;

(2) Sampling errors;

(3) Non-representativeness.

6.3.7.7 Work Sampling Systems

Work sampling system involves observing the workforce for a sufficient number of times at random intervals. A note is made and recorded after every observation, and this information helps to determine the proportion of time used by the workforce in the defined activities. Work sampling is done to test the ability of a worker and evaluate his performance. The steps involved to make up a work sampling system are as follows:

(1) The first step in the process of work sampling is to define the problem.

(2) State the main objectives of the problem.

(3) This is the time to describe each element in detail.

(4) Obtain the approval of the workers that are to be studied.

(5) Determine the desired accuracy of the final results in the form of a percentage.

(6) Design the observation form to see and record the data.

(7) Now decide on the preliminary sampling or the number of considerations that should be made.

(8) Determine the random timings that have been chosen for making the observations.

(9) The next step involves making instant observations at every visit irrespective of the fact that the workforce is sitting idle or is working.

(10) It is the time to determine the activity proportion in the works sample with the help of the formula. The formula is:

P = Number of times found working/Total number of observations made

(11) Determine the confidence level that is required for the work sampling.

(12) Find the Z value corresponding to the confidence level.

(13) It is the time to make an interval estimate of the proportion of working of the worker.

(14) Find the data at the end of each day.

(15) Check its precision or accuracy at the end of the study.

(16) Prepare a detailed report and state your findings and the actual result.

6.3.8 Predetermined Motion Time Systems (PMTS)

6.3.8.1 PMTS

Predetermined motion time systems (PMTS) are work measurement systems based on the analysis of work into basic human movements, classified according to the nature of each movement and the conditions under which it is made. Tables of data provide a time, at a defined rate of working, for each classification of each movement.

The first PMTS (since designated as "first-level" systems) were designed to provide times for detailed manual work and thus consisted of fundamental movements (reach, grasp, move, etc.) and associated times.

Large amounts of research, data collection, analysis, synthesis, and validation are required to produce PMTS data, and the number of such systems is very low. "Higher level" systems have since been devised, most commonly by combining these fundamental movements into common, simple manual tasks. Such higher-level systems are designed for faster standard setting of longer cycle activity.

Criticisms of PMTS relate to their inability to provide data for movements made under "unnatural" conditions (such as working in cramped conditions or with an unnatural body posture) or for mental processes and their difficulty in coping with work which is subject to interruptions. However, various systems have been derived for "office work", which include tasks with a simple and predictable mental content.

Conversely, one of the significant advantages of PMTS is that they require a detailed description of the working method-and are thus useful for studying how work is done (and how it can be improved) as well as measuring the time it should take.

Many PMTS are proprietary systems and users must either attend a designated and approved training course and/or pay a royalty for use of the data.

One of the major PMTS systems is MTM (Methods-Times Measurement) which is actually a "family" of systems operating at different levels and applicable to different types of work. MTM1—the "highest-level" or most detailed member of the family—was developed in the 1940s by analyzing large numbers of repetitive cycles of manual work on film. MTM gives values for such basic hand/arm motions as: Reach, Move, Turn, Grasp, Position, Disengage, and Release, together with a small set of full body motions. The time taken to Reach to an object is then given by a table based on the kind of Reach (e. g. , whether the object is in a fixed location—such as a tool in a tool holder—or is a single object located on a bench, or jumbled together with other objects, etc.) and the distance to be Reached. Similar tables give times for each of the other basic movements categorized and measured similarly. MTM is suitable for measuring short cycle, highly repetitive work. Other members of the MTM family use lower-level motions (so that in MTM2, for example, the MTM1 motions of Reach and Grasp are combined into a composite motion, GET). MTM2 is thus quicker to apply, but more suited to longer-cycle work where the fine level of discrimination of MTM1 is unnecessary in terms of meeting accuracy requirements.

6. 3. 8. 2 MODAPTS

MODAPTS (as shown in Figure 6-6) or modular arrangement of predetermined time standards, is a simple and descriptive language for understanding work, tasks, or activities. All MODAPTS times are multiples of 0. 129 seconds.

MODAPTS is a form of shorthand or a succinct language for describing the sequence of body actions involved in carrying out particular work.

6. 3. 9 Standard Data

6. 3. 9. 1 Meaning of Standard Data

Generally large number of man hours is spent in setting the time standards by using stop-watch

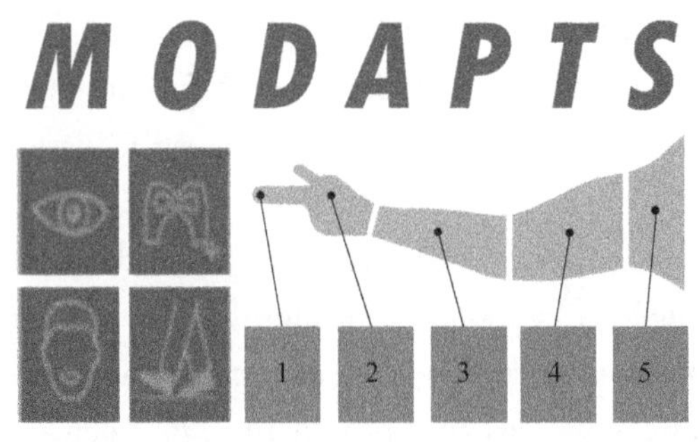

扫一扫查看彩图

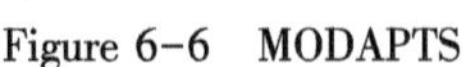

Figure 6-6 MODAPTS

time study. Further management is anxious to have the time standards before the jobs are actually manufactured for cost estimating, scheduling, planning and other decision-making purposes. In this case, advantage of previous time standards already on hand can be taken.

Every operation consists of number of small work elements which are repeated in various combinations. The time values for these small work elements are established accurately, and these values are used without further time study, whenever the element occurs. These stan-dardized timings for such elements are known as standard data.

For calculations of standard time, using the standard data, first step is to standardize the method (by method study). Then break the operation into small elements and note down their timings from standard data tables and then add the timing of all such elements to get the standard time of the operation.

6.3.9.2 Methods for Determining Standard Data

Following methods are commonly used for determining the Standard Data. These are also known as "Predetermined Motion Time Standard" (P. M. T. S.) systems.

(1) Work Factor (WF) system.

(2) Method Time Measurement (MTM).

(3) Basic Motion Time (BMT) study.

6.3.9.3 Uses of Standard Data

(1) Standard data helps in determining in advance that how long it will take to perform an operation in the shop.

(2) It helps in comparing two methods and determining best method.

(3) Evaluating proposed methods in advance of actual production.

(4) Helps in estimating the time required and labor cost.

(5) For checking standards established by time study.

6.3.9.4 Advantages of Standard Data

(1) Standard time can be found even when the product is not being manufactured.

(2) Provides a basis for decision making budgeting estimating etc.
(3) Various time study persons can arrive at identical time standards for given method.
(4) There can be no dispute about partiality or incorrectness of time study.
(5) Economical than the time study by a stop-watch method.
(6) Rating or efficiency of the operator can be determined correctly.

词汇

生词	音标	释义
sequential	[sɪˈkwenʃəl]	*adj.* 按次序的；顺序的
implementation	[ˌɪmplɪmenˈteɪʃən]	*n.* 实施；执行
sketch	[sketʃ]	*n.* 素描；速写；草图
synchronization	[ˌsɪŋkrənaɪˈzeɪʃ（ə）n]	*n.* 同时；同时性
consecutive	[kənˈsekjʊtɪv]	*adj.* 连续的
surveillance	[səːrˈveɪləns]	*n.* 监控；监视

长难句

A term used to embrace the techniques of method study and work measurement which are employed to ensure the best possible use of human and material resources in carrying out a specified activity.

（作业研究是）用于包含方法研究和作业测定技术的术语，用于确保在执行特定活动时尽可能最佳地使用人力和物力资源。

A procedure for examining the various activities associated with the problem which ensures a systematic, objective, and critical evaluation of the existing factors and in addition and imaginative approach while developing improvements.

（方法测定是）审查与问题有关的各种活动的程序，确保对现有因素进行系统、客观和批判性的评估，并在制定改进措施的同时采取富有想象力的方法。

7 Production Planning and Control

7.1 Introduction to Production Planning

7.1.1 Production & Production Models

Production is a process of combining various material inputs and immaterial inputs (plans, know-how) in order to make something for consumption (output). It is the act of creating an output, a good or service which has value and contributes to the utility of individuals[1].

A production model is a numerical description of the production process and is based on the prices and the quantities of inputs and outputs. There are two main approaches to operationalize the concept of production function. We can use mathematical formulae, which are typically used in macroeconomics (in growth accounting) or arithmetical models, which are typically used in microeconomics and management accounting.

We use here arithmetical models because they are like the models of management accounting, illustrative and easily understood and applied in practice. Furthermore, they are integrated to management accounting, which is a practical advantage. A major advantage of the arithmetical model is its capability to depict production function as a part of production process. Consequently, production function can be understood, measured, and examined as a part of production process.

There are different production models according to different interests. Here we use a production income model and a production analysis model in order to demonstrate production function as a phenomenon and a measurable quantity.

7.1.2 Production Planning

Production planning is the planning of production and manufacturing modules in a company or industry. It utilizes the resource allocation of activities of employees, materials and production capacity, in order to serve different customers. Production planning can be combined with production control into production planning and control, or it can be combined with enterprise resource planning. Figure 7-1 shows Framework of production planning Activities.

Production planning is the future of production. It can help in efficient manufacturing or setting up of a production site by facilitating required needs. A production plan is made periodically for a specific time period, called the planning horizon. It can comprise the following activities:

Determination of the required product mix and factory load to satisfy customers' needs.

Matching the required level of production to the existing resources.

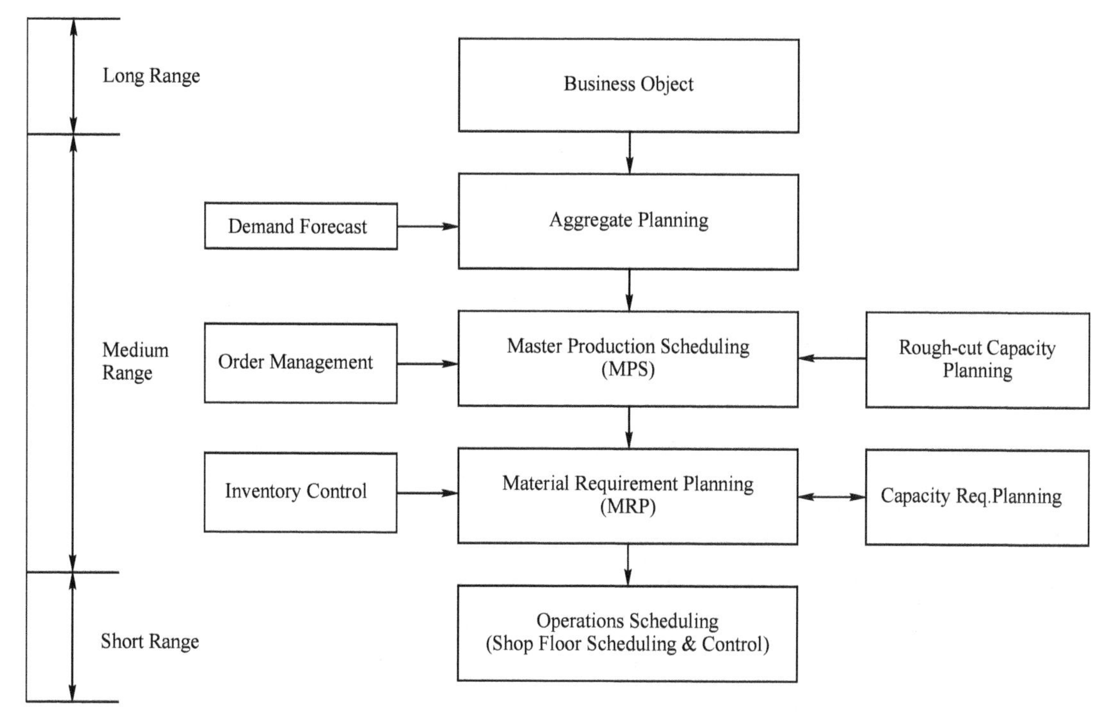

Figure 7-1 Framework of production planning activities

Scheduling and choosing the actual work to be started in the manufacturing facility.

Setting up and delivering production orders to production facilities.

In order to develop production plans, the production planner or production planning department needs to work closely together with the marketing department and sales department.

A critical factor in production planning is "the accurate estimation of the productive capacity of available resources, yet this is one of the most difficult tasks to perform well". Production planning should always take "into account material availability, resource availability and knowledge of future demand".

7.1.3 Types of Planning

Different types of production planning can be applied:

(1) Advanced planning and scheduling;

(2) Capacity planning;

(3) Master production schedule;

(4) Material requirements planning;

(5) MRP Ⅱ;

(6) Scheduling;

(7) Workflow.

Related kind of planning in organizations:

(1) Employee scheduling;

(2) Enterprise resource planning;
(3) Inventory control;
(4) Product planning;
(5) Project planning;
(6) Process planning, redirects to computer-aided process planning;
(7) Sales and operations planning;
(8) Strategy.

7.1.4 The Steps of Production Planning

(1) Preparation for production planning. First of all, predicting market demand in planning period, accounting company's own production capacity and providing an external needs and internal possibility basis.

(2) To determine the targets of the plan. Secondly, according to the principle that is making full use of various resources and increasing economic benefits to meet the social needs. On the basis of overall balance to determine targets of the production planning such as product variety, quality, yield, and value indicators which should be finished in the planning period.

(3) Account and balance the production capacity. Third, according to account production capacity and balance implementation of the planning task while make full use of the production capacity to achieve the best economic benefits.

(4) Determine product production schedule. Next is to arrange product production schedule properly, in which ensures the realization of production targets in time, and also ensures the production have a nice tie with marketing and balanced production, so that the production capacity is fully utilized and to reduce production costs.

(5) Organized and checked of the implementation of production planning. After the completion of production planning, the key to achieve prospective objectives is to organize implementation of the plan. During the implementation, checking the implementation in time, tracking the production process, analyzing and evaluating, summing up experience and to correct deviations are necessary. Based on above all works, the last thing is to table and fill in those production schedules[2].

7.1.5 Mode of Production

In the writings of Karl Marx and the Marxist theory of historical materialism, a mode of production is a specific combination of the following:

Productive forces: these include human labor power and means of production (e.g., tools, productive machinery, commercial and industrial buildings, other infrastructure, technical knowledge, materials, plants, animals, and exploitable land).

Social and technical relations of production: these include the property, power and control relations governing society's productive assets (often codified in law), cooperative work relations and forms of association, relations between people and the objects of their work and the relations between social classes.

Marx regarded productive ability and participation in social relations as two essential characteristics of social reproduction and that the particular modality of these relations in capitalist production are inherently in conflict with the increasing development of human productive capacities.

Modes of production include tribal and Neolithic modes of production; Asiatic mode of production; antique or ancient mode of production; feudal mode of production; capitalist mode of production; socialist mode of production; communist mode of production.

7.2 Planning and Control Technology

Different types of production methods, such as single item manufacturing, batch production, mass production, continuous production etc. have their own type of production planning.

7.2.1 Demand Forecasting

Demand forecasting is a field of predictive analytics which tries to understand and predict customer demand to optimize supply decisions by corporate supply chain and business management. Demand forecasting involves quantitative methods such as the use of data, and especially historical sales data, as well as statistical techniques from test markets. Demand forecasting may be used in production planning, inventory management, and at times in assessing future capacity requirements, or in making decisions on whether to enter a new market.

The methods of demand forecasting (as shown in Figure 7-2) include qualitative forecasting and quantitative forecasting.

7.2.2 Inventory Analysis and Control

In materials management, ABC analysis is an inventory categorization technique. ABC analysis divides an inventory into three categories— "A items" with very tight control and accurate records, "B items" with less tightly controlled and good records, and "C items" with the simplest controls possible and minimal records.

The ABC analysis provides a mechanism for identifying items that will have a significant impact on overall inventory cost, while also providing a mechanism for identifying different categories of stock that will require different management and controls.

The ABC analysis suggests that inventories of an organization are not of equal value. Thus, the inventory is grouped into three categories (A, B and C) in order of their estimated importance. "A" items are very important for an organization. Because of the high value of these "A" items, frequent value analysis is required. In addition to that, an organization needs to choose an appropriate order pattern (e.g., "just-in-time") to avoid excess capacity. "B" items are important, but of course less important than "A" items and more important than "C" items. Therefore, "B" items are intergroup items. "C" items are marginally important.

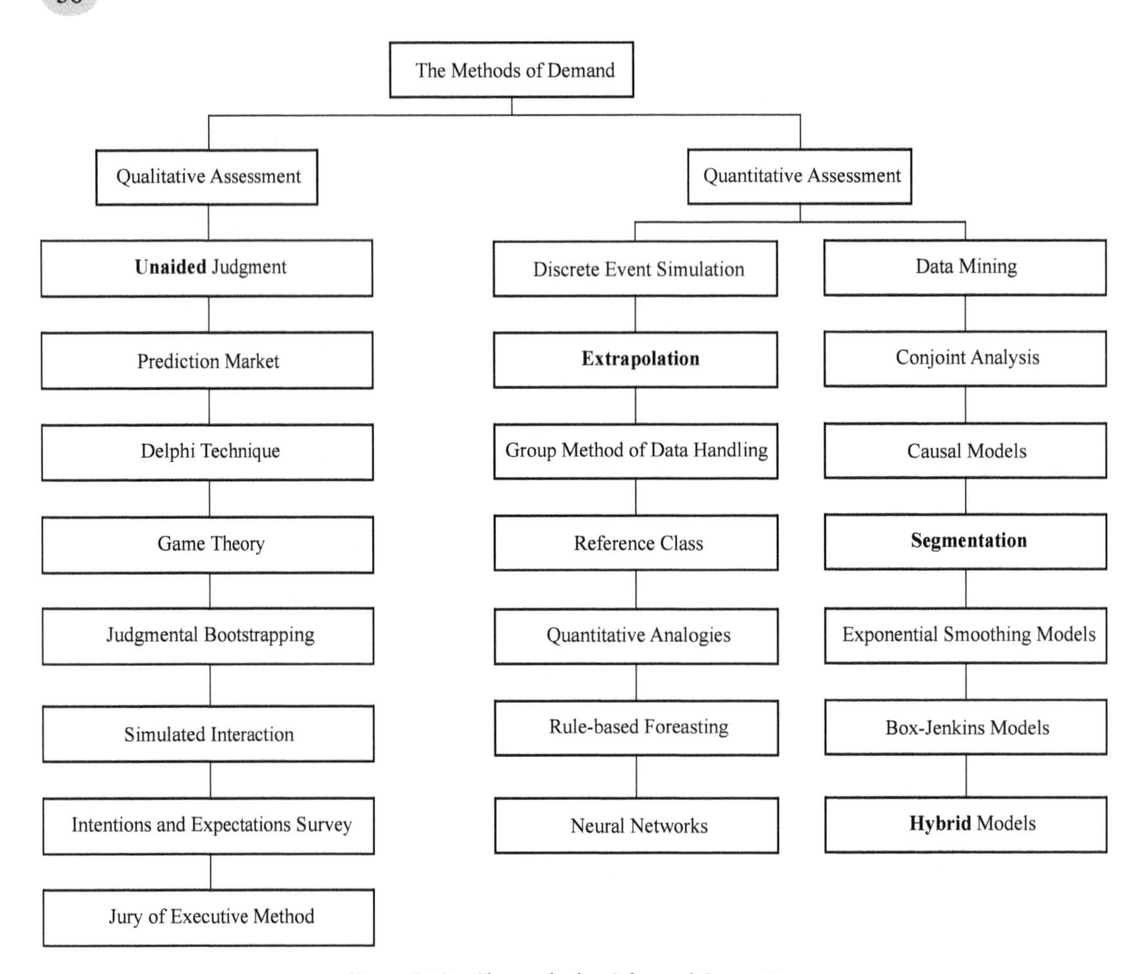

Figure 7-2 The methods of demand forecasting

There are no fixed thresholds for each class, and different proportions can be applied based on objectives and criteria. ABC Analysis is similar to the Pareto principle in that the "A" items will typically account for a large proportion of the overall value, but a small percentage of the number of items.

Examples of ABC class are:

(1) "A" items——20% of the items accounts for 70% of the annual consumption value of the items.

(2) "B" items——30% of the items accounts for 25% of the annual consumption value of the items.

(3) "C" items——50% of the items accounts for 5% of the annual consumption value of the items.

Another recommended breakdown of ABC classes:

(1) "A" approximately 10% of items or 66.6% of value;

(2) "B" approximately 20% of items or 23.3% of value;

(3) "C" approximately 70% of items or 10.1% of value.

7.2.3 MPS

A master production schedule (MPS) is a plan for individual commodities to be produced in each time period such as production, staffing, inventory, etc. It is usually linked to manufacturing where the plan indicates when and how much of each product will be demanded. This plan quantifies significant processes, parts, and other resources in order to optimize production, to identify bottlenecks, and to anticipate needs and completed goods. Since a MPS drives much factory activity, its accuracy and viability dramatically affect profitability. Typical MPSs are created by software with user tweaking.

Due to software limitations, but especially the intense work required by the "master production schedulers", schedules do not include every aspect of production, but only key elements that have proven their control effectivity, such as forecast demand, production costs, inventory costs, lead time, working hours, capacity, inventory levels, available storage, and parts supply. The choice of what to model varies among companies and factories. The MPS is a statement of what the company expects to produce and purchase (i.e., quantity to be produced, staffing levels, dates, available to promise, projected balance).

The MPS translates the customer demand (sales orders, PIR's), into a build plan using planned orders in a true component scheduling environment. Using MPS helps avoid shortages, costly expediting, last minute scheduling, and inefficient allocation of resources. Working with MPS allows businesses to consolidate planned parts, produce master schedules and forecasts for any level of the Bill of Material (BOM) for any type of part.

7.2.4 MRP

Material requirements planning (MRP) is a production planning, scheduling, and inventory control system used to manage manufacturing processes. Most MRP systems are software-based, but it is possible to conduct MRP by hand as well.

An MRP system is intended to simultaneously meet three objectives:

(1) Ensure raw materials are available for production and products are available for delivery to customers.

(2) Maintain the lowest possible material and product levels in store.

(3) Plan manufacturing activities, delivery schedules and purchasing activities.

7.2.5 CRP

Capacity planning is the process of determining the production capacity needed by an organization to meet changing demands for its products. In the context of capacity planning, design capacity is the maximum amount of work that an organization is capable of completing in a given period. Effective capacity is the maximum amount of work that an organization is capable of completing in a given period due to constraints such as quality problems, delays, material handling, etc.

The phrase is also used in business computing and information technology as a **synonym** for capacity management. IT capacity planning involves estimating the storage, computer hardware, software and connection infrastructure resources required over some future period of time. A common concern of enterprises is whether the required resources are in place to handle an increase in users or number of interactions. Capacity management is concerned about adding central processing units (CPUs), memory and storage to a physical or virtual server. This has been the traditional and vertical way of **scaling up** web applications, however IT capacity planning has been developed with the goal of forecasting the requirements for this vertical scaling approach.

A discrepancy between the capacity of an organization and the demands of its customers results in inefficiency, either in under-utilized resources or unfulfilled customer demand. The goal of capacity planning is to minimize this discrepancy. Demand for an organization's capacity varies based on changes in production output, such as increasing or decreasing the production quantity of an existing product or producing new products. Better utilization of existing capacity can be accomplished through improvements in overall equipment effectiveness (OEE). Capacity can be increased through introducing new techniques, equipment and materials, increasing the number of workers or machines, increasing the number of shifts, or acquiring additional production facilities.

Capacity is calculated as (number of machines or workers) × (number of shifts) × (utilization) × (efficiency).

7.2.6 Production Control

Production control is the activity of monitoring and controlling a large physical facility or physically **dispersed** service. It is a "set of actions and decision taken during production to regulate output and obtain reasonable assurance that the specification will be met." The American Production and Inventory Control Society, nowadays APICS, defined production control in 1959 as:

Production control is the task of predicting, planning and scheduling work, considering manpower, materials availability and other capacity restrictions, and cost so as to achieve proper quality and quantity at the time it is needed and then following up the schedule to see that the plan is carried out, using whatever systems have proven satisfactory for the purpose.

Production planning and control in larger factories is often run from a production planning department run by production controllers and a production control manager. Production monitoring and control of larger operations is often run from a central space, called a control room or operations room or operations control center (OCC).

The emerging area of Project Production Management (PPM), based on viewing project activities as a production system, adopts the same notion of production control to take steps to regulate the behavior of a production system where in this case the production system is a capital project, rather than a physical facility or a physically dispersed service.

Production control is to be contrasted with project controls. As explained, project controls have developed to become centralized functions to track project progress and identify deviations from plan and to forecast future progress, using metrics rooted in accounting principles [3].

7.2.7 Project Planning and Controlling

Project planning is part of project management, which relates to the use of schedules such as Gantt charts to plan and subsequently report progress within the project environment. Project planning can be done manually or by the use of project management software.

Initially, the project scope is defined and the appropriate methods for completing the project are determined. Following this step, the durations for the various tasks necessary to complete the work are listed and grouped into a work breakdown structure. Project planning is often used to organize different areas of a project, including project plans, workloads and the management of teams and individuals. The logical dependencies between tasks are defined using an activity network diagram that enables identification of the critical path. Project planning is inherently uncertain as it must be done before the project is actually started. Therefore, the duration of the tasks is often estimated through a weighted average of optimistic, normal, and pessimistic cases. The critical chain method adds "buffers" in the planning to anticipate potential delays in project execution. Float or slack time in the schedule can be calculated using project management software. Then the necessary resources can be estimated and costs for each activity can be allocated to each resource, giving the total project cost. At this stage, the project schedule may be optimized to achieve the appropriate balance between resource usage and project duration to comply with the project objectives. Once established and agreed, the project schedule becomes what is known as the baseline schedule. Progress will be measured against the baseline schedule throughout the life of the project. Analyzing progress compared to the baseline schedule is known as earned value management.

The inputs of the project planning phase 2 include the project charter and the concept **proposal**. The outputs of the project planning phase include the project requirements, the project schedule, and the project management plan.

词汇

生词	音标	释义
production	[prəˈdʌkʃən]	*n.* 成果；产品；生产；作品
immaterial	[ˌɪməˈtɪərɪəl]	*adj.* 非物质的；无形的；不重要的；非实质的
numerical	[nuːˈmerɪkl]	*adj.* 数值的；数字的；用数字表示的（等于 numeric）
operationalize	[ˌɒpəˈreɪʃənəlaɪz]	*vt.* 使开始运转；实施；使用于操作
macroeconomics	[ˌmækrəʊˌiːkəˈnɒmɪks]	*n.* 宏观经济学；[经] 宏观经济学的
microeconomics	[ˌmaɪkrəʊiːkəˈnɒmɪks]	*n.* 微观经济学
arithmetical	[ˌærɪθˈmetɪkl]	*adj.* 算术的；算术上的
illustrative	[ɪˈlʌstrətɪv]	*adj.* 说明的；作例证的；解说的
enterprise	[ˈentəpraɪz]	*n.* 企业；事业；进取心；事业心

computer-aided		*adj.* [计] 计算机辅助的，电脑辅助
implementation	[ˌɪmplɪmenˈteɪʃən]	*n.* [计] 实现；履行；安装启用
deviation	[ˌdiːvɪˈeɪʃən]	*n.* 偏差；误差；背离
infrastructure	[ˈɪnfrəstrʌktʃə]	*n.* 基础设施；公共建设；下部构造
inherently	[ɪnˈherəntlɪ, ɪnˈhɪrəntlɪ]	*adv.* 内在地；固有地；天性地
tribal	[ˈtraɪbl]	*adj.* 部落的；种族的
Neolithic	[ˌniːəˈlɪθɪk]	*adj.* [古] 新石器时代的；早先的
Asiatic	[ˌeɪsɪˈætɪk]	*n.* 亚洲人
		adj. 亚洲的；亚洲人的
Antique	[ænˈtiːk]	*adj.* 古老的，年代久远的；过时的，古董的；古风的，古式的
		n. 古董，古玩；古风，古希腊和古罗马艺术风格
		vi. 觅购古玩
feudal	[ˈfjuːdl]	*adj.* 封建制度的；领地的；世仇的
capitalist	[ˈkæpɪtəlɪst]	*n.* 资本家；资本主义者，资本主义拥护者
		adj. 资本主义的；资本家的
unaided	[ʌnˈeɪdɪd]	*adj.* 未受协助的；无助的
bootstrapping	[ˈbuːtstræpɪŋ]	*n.* 自举电路；引导指令；自展
		v. 依靠自己的努力获得成功；自展（bootstrap 的 ing 形式）
extrapolation	[ɪkˌstræpəˈleɪʃn]	*n.* [数] 外推法；推断
segmentation	[ˌsegmenˈteɪʃn]	*n.* 分割；割断；细胞分裂
hybrid	[ˈhaɪbrɪd]	*n.* 杂种，混血儿；混合物；
		adj. 混合的；杂种的
marginally	[ˈmɑːdʒɪnəli]	*adv.* 稍微，略微地；轻微地，很少的；边缘地；在页边；最低限度地
commodities	[kəˈmɒdɪtɪs]	*n.* 商品（commodity 的数）；日用品；商品期货
bottleneck	[ˈbɒtlˌnek]	*n.* 瓶颈；障碍物
allocation	[ˌæləˈkeɪʃn]	*n.* 分配，配置；安置
synonym	[ˈsɪnənɪm]	*n.* 同义词；同义字
scaling up		按比例放大；按比例增加
discrepancy	[dɪˈskrepənsɪ]	*n.* 不符；矛盾；相差
dispersed	[dɪˈspɜːst]	*adj.* 散布的；被分散的；被驱散的；
		v. 分散；传播（disperse 的过去分词）
manpower	[ˈmænˌpaʊə]	*n.* 人力；人力资源；劳动力
notion	[ˈnəʊʃn]	*n.* 概念；见解；打算

metrics	[ˈmetrɪks]	*n.* 度量；作诗法；韵律学
thresholds	[θˈreʃhəʊldz]	*n.* 阈值；[建] 门槛；临界值（threshold 的复数）
anticipate	[ænˈtɪsɪpeɪt]	*v.* 预料，预期；预见，预计（并做准备）；期盼，期望；先于……做，早于……行动；在期限内履行（义务），偿还（债务）；提前使用
pessimistic	[ˌpesɪˈmɪstɪk]	*adj.* 悲观的，厌世的；悲观主义的
buffers	[ˈbʌfəz]	*n.* DOS 命令
anticipate	[ænˈtɪsɪpeɪt]	*v.* 预料，预期；预见，预计（并做准备）；期盼，期望；先于……做，早于……行动；在期限内履行（义务），偿还（债务）；提前使用
baseline	[ˈbeɪslaɪn]	*n.* 基线；底线
proposal	[prəˈpoʊzl]	*n.* 提议，建议；求婚

长难句

Production is a process of combining various material inputs and immaterial inputs (plans, know-how) in order to make something for consumption (output). It is the act of creating an output, a good or service which has value and contributes to the utility of individuals.

生产是将各种物料投入和非物料投入（计划、技术诀窍）结合起来，以制造某种消费品（产出）的过程。它是创造一种产出、一种商品或服务的行为，这种产出、商品或服务具有价值并有助于个体的效用。

A production model is a numerical description of the production process and is based on the prices and the quantities of inputs and outputs.

生产模型是对生产过程的数值模拟，它是以投入和产出的价格和数量为基础的。

Production planning is the planning of production and manufacturing modules in a company or industry. It utilizes the resource allocation of activities of employees, materials, and production capacity, in order to serve different customers.

生产计划是一个公司或行业中生产和制造模块的计划。它利用员工、物料、生产能力等活动的资源配置，为不同的客户服务。

Productive forces: these include human labor power and means of production (e. g., tools, productive machinery, commercial and industrial buildings, other infrastructure, technical knowledge, materials, plants, animals, and exploitable land).

生产力：包括人力和生产资料（如工具、生产机械、商业和工业建筑、其他基础设施、技术知识、材料、植物、动物和可开发土地）。

Demand forecasting is a field of predictive analytics which tries to understand and predict customer demand to optimize supply decisions by corporate supply chain and business management.

需求预测是预测分析的一个领域，它试图通过企业供应链和企业管理来理解和预测客户需求，从而优化供应决策。

A master production schedule (MPS) is a plan for individual commodities to be produced in each time period such as production, staffing, inventory, etc.

主生产计划（MPS）是在每个时间段内生产的单个商品的计划，如生产、人员配备、库存等。

Capacity planning is the process of determining the production capacity needed by an organization to meet changing demands for its products.

产能规划是确定一个组织为满足产品不断变化需求所需生产能力的过程。

Reference

[1] Kotler P, Armstrong G, Brown L, et al. Pearson Education [M]. 7th Ed. Australia: Prentice Hall, 2006.

[2] 周跃进，任秉银. 工业工程专业英语 [M]. 北京：机械工业出版社，2017.

[3] Arbulu R J, Choo H J, Williams M. Proc. International Conference on Innovative Production and Construction (IPC 2016) [P]. Australia: 3-5 October 2016.

8 Logistics Engineering

8.1 Introduction of Logistics Engineering

Logistics is concerned with the total movement of materials through the **enterprise**, including the movements of documents and other facilitators to movement. It includes the management of the interruptions to movement, such as **storage**, if storage is necessary, to the efficiency of the production process. Logistics can only be successfully undertaken as an integrated activity in the business environment if it is allied to information systems. In addition, manufacturing and retail companies have been able to contract out parts of their logistics operation to third parties like distribution companies, while still retaining effective control by the use of accurate and timely information transfer.

Logistics involves functions of customer service, demand forecasting, order processing, production planning, packaging, storage, **inventory** management, transportation, and return goods handling and so on[1].

The institute of logistics gave the following definition: "**Logistics is the time related positioning of resource or the strategic management of the total supply chain. The supply chain is a sequence of events intended to satisfy a customer. It can include procurement, manufacture, distribution, and waste disposal, together with associated transport, storage and information technology.**"

Logistics Engineering means the management process of choosing the best scheme under the guidance of theories about system engineering and planning, managing, controlling the system with lowest cost, high efficiency, and good customer service for the purpose of improving economy profits of the society and enterprises[2].

8.2 Logistics: Origins and Evolution

The word "logistics" has two origins: **mathematical and military**. Figure 8-1 shows two origins of logistics.

8.2.1 Mathematical Origin

The origin of the word logistics is Greek: logistikos, which corresponds to mathematical reasoning. This term was first used by the Greek philosopher Plato (428-448 BC) and was the origin of the Latin word logisticus. It was first used in the French language in 1590 as an adjective to describe a logical reasoning.

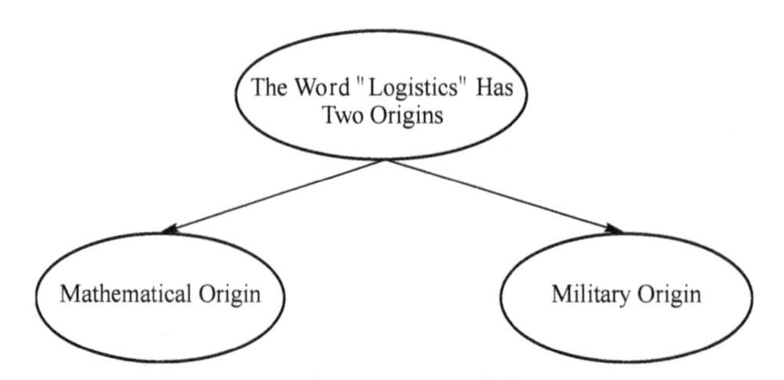

Figure 8-1 Two origins of logistics

At the start of the 20th Century, Bertrand Russell, British mathematician, and logistician (1872-1970), highlighted the close link between logistics and mathematical logistics. This represented the beginning of the theory of algorithms. In this sense, logistics refers to the art of organizing a calculation in stages to attain an aim. This discipline is called algorithmic logics.

8.2.2 Military Origin

The word "logistics" equally finds its source in the battlefields. Its meaning derives from the rank of an officer in charge of the dwellings of the troops during combat. Since ancient antiquity, logistics has played an important role in military activity. However, it was in the 20th Century, during the First and Second World Wars, that there have been important developments concerning the reflection in the practice of the subject, hence the development of a recognized science, called military logistics. The aim of this science is to regulate human resources, food and material flows in order to ensure, among others, the support in foodstuffs and the provision of equipment and transport means for the armed forces. The term "logistics" indicates, by an abuse of language, the science of the plan and execution of the transfer of armed forces and their maintenance.

8.2.3 Evolution

Logistics is an old word that has been used for centuries. However, despite the fact that the term has evolved with time, it always refers to activities which aim to select the adequate resources (human, material, informational, etc.) to provide a service at the lowest cost.

Considered for a long time as the "logistics of modern times", logistics is today a general concept concerning all the operations that determine the movement of products. Many methodologies and organization methods have derived from this concept, the best-known being "Supply Chain Management" (SCM) which marked the beginning of the 21st Century with its massive use by the professional and academic worlds. Basically, it deals with optimizing the LC in an organization. According to the Association Francoise pour la Logistique1 (ASLOG), the aim of SCM is to coordinate the management of a network in terms of costs, deadlines, and quality from the supplier's providers up to the distribution of these same goods to the final consumers.

In the 1940s, logistics mainly dealt with the functions related to the physical flows of distribution. However, owing to industrial development, the concept has evolved throughout the years, now including the evolution of internal and external factors to the firms.

Three great periods have marked the history of logistics regarding its incorporation in Information Systems (IS): IS of different logistics components (products, suppliers, customers, business orders, etc.). Thus, IS are classified into three types: separated (before 1975), integrated (after 1975) or cooperative (from the 1990s onwards). The presence of a logistics manager has become essential since the emergence of integrated IS:

Separated logistics (service logistics): separation of all the logistics components.

Integrated logistics (function logistics): integration of all the logistics components into a system. All the actors of this system communicate and possess a global view of the impact of logistics over the entire organization.

Cooperative logistics (process logistics): the cooperation between the different logistics actors becomes important because of the insufficiency to integrate all the components under the same system. Cooperation creates real partnerships between suppliers, customers, competitors, etc., and participates in the emergence of communicative logistics tools such as Enterprise Resource Planning (ERP). The latter deals simultaneously with the organization's internal and external data in order to better manage resources. Communication is done through the Exchange of Digitized Data.

8.3 Related Concepts of Logistics Engineering

8.3.1 Logistics Process

Cooperative logistics process is shown as Figure 8-2.

Figure 8-2 Cooperative logistics process

A process is an ordered sequence of operations, tasks, actions, or activities responding to a certain diagram and leading to a result.

A Business Process, equally called operational process, involves business rules. These rules are the sources for decision (to perform the activities or not) at the heart of the process. Business Process activities are performed by actors (human or automatic) with the help of adapted means. The performance of each Business Process activity contributes to the achievement of an expected result. The business character of the process is equally expressed by the nature of the result, which must be effective, and should make sense to a final beneficiary. Nowadays, we talk of Workflow or Business Process Management, which corresponds to the electronic management of BPs.

A logistic process is a business process instance which aims to produce a result at the right moment, at the lowest cost and with the best possible quality service. It constitutes an operation and an event sequence which resorts to a set of means (human, material, and informational resources) and controls the corresponding flows in order to achieve an objective at the lowest cost. These means can be human (staff), material (installations and equipment) or methodological (techniques and methods) and must be efficiently coordinated in order to attain the best possible service level. The logistic process crosses traditional organizational boundaries between the firms and their functions.

The X50-600 regulation of the association francoise de la normalisation 3 (AFNOR) defines three **deployment phases for a logistic process**:

(1) **Planning**: identifying the needs, conceiving the LS, and developing it.

(2) **Execution**: for the implementation of the conceived system and its development during the planning phase. This implementation must produce the desired result at the lowest cost while adhering to the deadlines.

(3) **Mastery**: for flow control in order to identify the possible adjustments to be made at the previous phases, so that the system may continue to produce what it must, at the lowest cost while adhering to the deadlines. Figure 8-3 shows three deployment phases for a Logistic Process.

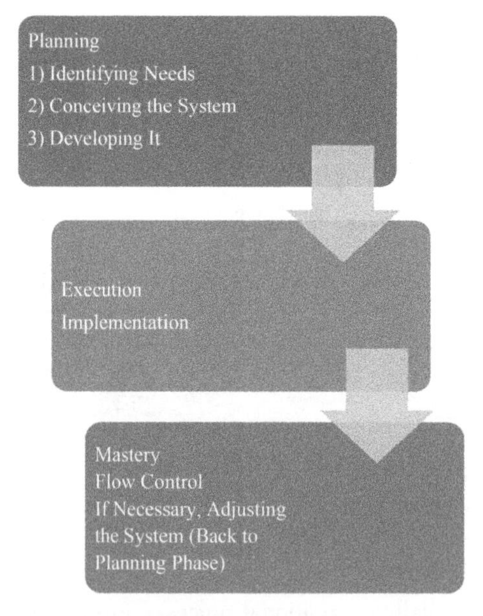

Figure 8-3 Three deployment phases for a logistic process

扫一扫查看彩图

8.3.2 Logistics System

A production and logistics system is an integrated system consisting of various entities that work together in order to acquire raw materials, convert these raw materials into specified final products, and deliver these final products to markets.

The production section of these systems includes the design and management of the entire manufacturing process. The logistics section determines how raw materials are sourced and transported from suppliers to plants, how products are **retrieved** and delivered from plants to distribution centers/warehouses to retailers, as well as how return **commodities** are transported back (reverse logistics). The typical production and logistics system consists of: suppliers of components and raw materials; manufacturing plants; zero, one, or more distribution echelons with distribution centers/warehouses; customers; recycling centers for used products and returned packaging containers; transportation channels that link all of the above components.

Over the last decades, competitive pressures have posed the challenge of simultaneously prioritizing the four dimensions of competition: flexibility, cost, quality, and timely delivery. In addition to these dimensions, other factors such as the speed with which products are designed, produced, and distributed, as well as the need for higher efficiency and lower operational costs, are forcing companies to continuously search for ways to improve their supply chain operations. Deterministic or stochastic optimization models and algorithms, decision support systems, and to computerized analysis tools are used to improve the performance of the production and logistics systems of companies in order to remain competitive under the threat of increasing competition. The research done in this area within Sabanci University primarily addresses the production and logistics systems planning problems using mathematical programming models, in addition to developing heuristics. Some of the studied problems include but are not restricted to the scheduling of jobs in a production facility, transportation problems, vehicle routing, production and distribution planning and the optimization of container terminal operations.

(1) Material handling system.

Material handling systems mean the control of materials and products for project use in various stages starting from manufacturing, storage, distribution, consumption, and finally disposal. The system must ensure the safe handling of all project materials. In this process, the material handling system uses various manual, automatic or semi-automatic equipment known as material handling equipment. So, the material handling system basically deals with the safety of material handling equipment & their operations.

Material Handling Systems are very important mechanisms in supply chain management as it efficiently manages the material movement in a controlled way. The material handling operation varies between manufacturing, storage, construction, and transportation based on industry types.

The main objective of the material handling system is to ensure proper handling, lifting & offloading equipment in order to ensure a safe workplace. It ensures the operations are in line with the required guidelines & project specifications. Debating with these standards and guidelines means tolerating the life of self and stakeholders. Basically, standard guidelines are followed in the construction of refineries, chemical & petrochemical, pharmaceutical companies, and power plants[5].

(2) Warehouse management system.

A Warehouse Management System (WMS) consists of software and processes that allow organizations to control and administer warehouse operations from the time goods or materials enter a warehouse until they move out.

Warehouses sit at the center of manufacturing and supply chain operations because they hold all of the material used or produced in those processes, from raw materials to finished goods. The purpose of a WMS is to help ensure that goods and materials move through warehouses in the most efficient and cost-effective way. A WMS handles many functions that enable these movements, including inventory tracking, picking, receiving, and put away. A WMS also provides visibility into an organization's inventory at any time and location, whether in a facility or in transit.

Although a WMS is complex and expensive to implement and run, organizations gain numerous benefits that can justify the complexity and costs.

Implementing a WMS can help an organization reduce labor costs, improve inventory accuracy, improve flexibility and responsiveness, decrease errors in picking and shipping goods, and improve customer service. Modern warehouse management systems operate with real-time data, allowing the organization to manage the most current information on activities like orders, shipments, receipts, and any movement of goods.

(3) Cold chain logistics system.

Cold chain management includes all of the means used to ensure a constant temperature (between +2°C and +8°C) for a product that is not heat stable (such as vaccines, serums, tests, etc.), from the time it is manufactured until the time it is used.

The cold chain must never be broken. Vaccines are sensitive to heat and extreme cold and must be kept at the correct temperature at all times.

Health workers at all levels are often responsible for maintaining the cold chain while vaccines are stored in the vaccine stores at the province and county levels, or while they are being transported to township and villages, and while they are being used during immunization sessions or rounds. More and more often it is becoming the logistician's responsibility to manage the cold chain as a part of the supply chain.

The logistics staff must be trained to both use and manage these materials. This includes having appropriate and efficient logistics mechanisms to manage shipping, fuel, spare parts etc. Without training, the program will be seriously compromised and put at risk.

(4) Reverse logistics system.

While conventional logistics optimizes the flow of goods from producer to consumer, reverse logistics manages the processes for inverting that flow to deal with returned parts, materials, and products from the consumer back to the producer. Most often, this includes warranty recovery, value recovery, repair, redistribution, product recalls, used parts and replacement materials for refurbishment, service or product contract returns, and end-of-life recycling.

With the purpose of optimizing supply chain efficiency and asset recovery rates, applying a reverse logistics system has increasingly become a tool that positively impacts profitability as well as assisting an organization in meeting sustainability goals. With the growth of sustainability initiatives, more companies have adopted the use of recycled materials in production and have developed procedures for the responsible disposal of products that cannot be recycled or reused. For instance, a growing number of cell phone manufacturers have established procedures in place for consumers who wish to return an older model and ensure that the device is refurbished or recycled rather than dumped into the local landfill.

Thus, reverse logistics management has developed into a discipline that produces cost reductions, adds efficiencies, and improves the consumer experience. Producers have discovered value within returned assets and the benefits of streamlining repair, return and product reallocation processes.

8.3.3 Logistics Flow

A Logistics Flow (LF) corresponds to any matter able to circulate throughout the components of a LS. Flow circulation can be made bottom-up to top-down or the other way round. We distinguish three categories of LF:

(1) Physical flows: raw materials and finished products in the manufacturing process. The human flows can be considered as physical flows. Flow direction:

1) Usual logistics: bottom-up→top-down;

2) Reversed logistics: top-down→bottom-up.

(2) Information flows: the exchanged information between the actors of a LS. Flow direction: both directions are possible

1) Bottom-up→top-down (e.g., acknowledgement of receipt of an order);

2) Top-down→bottom-up (e.g., quote).

(3) Financial flows associated with physical flows. Flow direction:

1) Usual logistics: bottom-up → top-down;

2) Refunds: top-down → bottom-up.

In addition, these flows can be push or pull.

Push flows

Push flows refer to the top-down shipment of resources, from their reception at upstream zones (bottom-up), that is to say, from the supplier to the final customer. In this case, supplies are "pushed" the furthest toward the consumer. In military contexts, push flows correspond to the delivery of a resource considered enough for the field operator to provide.

Pull flows

Pull flows refer to a direct answer to a supply request, that is to say, to send the available resources when downstream (top-down) zones make the request[3]. Figure 8-4 shows logistics flow processes.

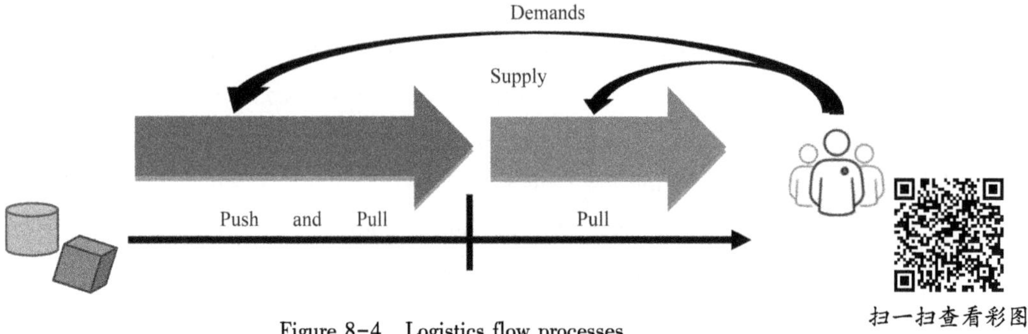

Figure 8-4 Logistics flow processes

8.3.4 Logistics Technology

(1) Transportation technology.

When you think of life-changing technological innovations in transportation, what comes to mind? Henry Ford's Model T? Commercial airlines? What about hybrid cars? All those answers would be correct. The point is these technologies completely upended the transportation sector's status quo and have had lasting impacts on how we get around.

Innovations in transportation technology are essentially born out of three necessities: efficiency, ease, and safety. Scientists and transportation industry professionals work side-by-side to ensure that these new technologies get more people (or things) to their destination faster, safer and with the fewest amount of resources possible. For example, this is why we've seen a shift away from coal-powered trains towards ultra-fast bullet trains, luxurious aircrafts to budget-friendly, cost-saving models and a switch from gas guzzling vehicles to 100% electric cars.

As technologies like AI, data science, manufacturing and deep learning become more advanced, so too will vehicles themselves. These fields act as the backbone for everything from autonomous vehicles to aerospace travel, and even function as the basis for transportation platforms like Uber and Lyft. Because of the enormous potential these technologies hold, transportation technology has become one of the fastest growing and highly contested fields in the world. Thousands of startups are racing to create the "next big thing" in the world of transportation.

The types of transportation technology:

1) Hyperloops;

2) Underground tunneling;

3) Aerospace;

4) Autonomous vehicles;

5) Last-mile robots;

6) Electric vehicles.

(2) Inventory technology.

Inventory management is the supervision of non-capitalized assets, or inventory, and stock items. As a component of supply chain management, inventory management supervises the flow of goods from manufacturers to warehouses and from these facilities to point of sale. A key function of inventory management is to keep a detailed record of each new or returned product as it enters or leaves a warehouse or point of sale.

Organizations from small to large businesses can make use of inventory management to manage their flow of goods. There are numerous inventory management techniques and using the correct one can lead to providing the correct goods, at the correct amount, place, and time.

Inventory control is a separate area of inventory management that is concerned with minimizing the total cost of inventory while maximizing the ability to provide customers with products in a timely manner. In some countries, the two terms are used as synonyms.

(3) Handling technology.

Investments in materials handling technology and equipment offer the potential for substantially improved logistics productivity. Materials handling processes and technologies impact productivity by influencing personnel, space, and capital equipment requirements. Material handling is a key logistics activity that can't be overlooked.

Logistical materials handling is concentrated in and around the warehouse. Specialized handling equipment is required for bulk unloading, such as for solids, fluids, pellets, or gaseous materials. Bulk handling of such material is generally completed using pipelines or conveyors. Handling systems can be classified as mechanized, semi - automated, automated, and information - directed[2].

Handling technology contains the tools of grippers, rotary grippers, swivel and rotary modules, robot accessories, linear modules, vacuum components, cutting tongs, separators, ball joints, adjustment jaws, sensors and holders, accessories.

(4) Packaging technology.

Packaging technology is the link between intralogistics and external logistics.

A brief definition of the term: Through single or multiple layers of packaging material, packaging technology makes products transportable. It is individually tailored to individual goods and modes of transport. Packaging technology always works according to the principle "as much packaging as necessary and as little as possible".

A detailed definition of packaging technology is provided by the website study engineering: "Packaging must not only protect the product inside and make them fit for transport but must also fulfill additional functions: They must be intelligently constructed, functionally designed and adequately printed. In addition, it should contain information about the product, make products sortable and be environmentally friendly."

Packaging technology is omnipresent. It is used for the protection, portioning, storage, transport, and marketing of goods. Packaging technology is especially important for logistics in that it represents the final step in intralogistics and is at the threshold of extra logistics.

(5) Containerization technology.

Containerization, method of transporting freight by placing it in large containers. Containerization is an important cargo-moving technique developed in the 20th century. Road-and-rail containers, sealed boxes of standard sizes, were used early in the century; but it was not until the 1960s that containerization became a major element in ocean shipping, made possible by new ships specifically designed for container carrying. Large and fast, container ships carry containers above deck as well as below; and their cargoes are easily loaded and unloaded, making possible more frequent trips and minimum lost time in port. Port facilities for rapid handling of containers are necessarily complex and expensive and usually justified only if there is large cargo traffic flowing both ways. A container may leave a factory by truck and be transferred to a railroad car, thence to a ship, and, finally, to a barge; such transfers of a containerized cargo would add substantially to cost.

(6) Logistics information technology.

The business of logistics is always evolving. Every day there are new supply chain innovations and technologies that are making the supply chain more efficient, responsive, and competitive. Here are the technologies to watch out for:

1) Robotics.

The introduction of robotics into the supply chain has reduced costs and improved efficiencies, while increasing productivity and accuracy. Today, robotics technologies take many forms and serve a number of important functions in the supply chain, from assembling widgets to reading barcodes to moving products from one area of a warehouse to another.

2) Internet of Things (IoT).

IoT is revolutionizing the supply chain, through the ability to embed sensors on parts, packages, and equipment to track them throughout their journey. This lets retailers know exactly where their goods are, from manufacturing to transit all the way to the warehouse and then to the consumer. Gartner predicts that by 2020 some element of IoT will be incorporated into more than half of new business processes.

3) Block chain.

Two major issues in supply chains today are their complexity and the lack of transparency within them. Blockchain—a distributed, digital ledger primarily used for cryptocurrency—has the potential to change all that. With blockchain, every transaction is recorded on a block, and each block links to each other. Within the supply chain, this would provide end-to-end transparency from manufacturing all the way to delivery.

4) On-demand warehousing.

On-demand warehousing has been recognized by gartner and featured in publications like Bloomberg and the wall street journal. It gives major retailers like walmart and ace hardware a scalable, flexible fulfillment and warehousing solution through a marketplace model and transactional pricing.

8.3.5 Logistics Management

Logistics management is the governance of supply chain management functions that helps organizations plan, manage, and implement processes to move and store goods.

Logistics management activities typically include inbound and outbound transportation management, fleet management, warehousing, materials handling, order fulfillment, logistics network design, inventory control, supply/demand planning and management of third - party logistics services providers.

8.3.5.1 Warehouse Inventory Management

(1) What is warehouse inventory management?

Warehouse inventory management is the process by which stock stored in a warehouse or other storage facility is received, tracked, audited, and managed for order fulfilment. Warehouse management also includes the replenishment of stock when predetermined minimum quantities are reached, refreshing your stock to optimal levels based on historical sales data. Much like broader inventory management processes, warehouse management is focused on managing incoming and outgoing products, all the while knowing where individual pieces are located.

Warehouse management is a specific subsection of a broader inventory management plan, which governs all products held by a company from the point of creating purchase orders for suppliers to ensuring the safe delivery of products to customers. Warehouse management is focused on the organization and tracking of stock while it is in storage, as well as how quickly certain items are sold.

(2) What is the difference between warehouse management and inventory management?

Warehouse management is specifically related to goods that are stored in warehouses and storage facilities, rather than those kept in storefronts or those that are used in the manufacturing process. It is part of the larger inventory management process, which monitors stock from the point of acquisition to the point of sale. But while that stock is in storage at your warehouse, you need a process in place to ensure it doesn't go missing, so it is ready to go when the time comes to sell it.

Warehouse management relates to a broader inventory management process by ensuring that items are shipped out to storefronts or customers in a timely manner. When a sale is made or a transfer order comes in, the warehouse should be set up to enable employees to quickly pick items, pack them and ship them. This means storing items in predictable locations and then tracking them as they move out the warehouse door through final delivery.

(3) How do warehouse inventory management systems work?

Warehouse inventory management software offers several key features to help you monitor the goods within your storage facilities and oversee inventory control. In some cases, warehouse management software is built into broader enterprise resource planning (ERP) software solutions; in other cases, warehouse management software serves as a stand - alone system. It is best to purchase a seamlessly integrated process if you want to manage your inventory across the entire ecosystem of your company.

(4) Best practices for managing your warehouse inventory.

1) Appoint a warehouse manager.

Running an efficient warehouse starts with appointing someone capable to lead; your business should recruit a warehouse manager who has: extensive experience operating a warehouse similar to the type you will be running.

2) Determine the warehouse layout.

The physical layout of your warehouse will either help or hinder your warehouse employees in quickly picking, packing, and shipping items out when a sale is made, or a transfer order is placed. According to Holton, separating warehouses into zones or lots and numbering aisles and bins can help warehouse workers navigate the storage facility more effectively.

Not every warehouse is set up the same way, but an organized warehouse is a prerequisite to efficient operations. How you design your warehouse space could vary depending on what types of products you store. For example, a warehouse storing large machinery might have specific zones but is unlikely to have bins and aisles, like a warehouse storing smaller retail products.

3) Establish a workflow.

With a leader appointed to monitor the operations of your warehouse and a system of organization in place, you will also need to put in place a specific workflow. The warehouse manager should have experience in this area, so work with them closely on how to establish a warehouse workflow that makes sense for your business.

4) Implement warehouse inventory management software.

A warehouse inventory management software can help automate and simplify a number of warehouse management tasks, as well as update a record of all existing stock in real time. As long as your warehouse team properly scans and catalogs items as they come into your warehouse and move throughout it, your warehouse inventory management software will reflect all your existing stock and its specific location in the warehouse.

8.3.5.2 Transportation and Distribution Management

Moving products from manufacturing plants to warehouses, between facilities and to distributors, can represent more than half of your total logistics costs. Add international sourcing or distribution, and those costs can skyrocket even higher.

When you work with NWCC for transportation, End to End solutions there are simply fewer vendors and tasks to manage. We can provide services from a total transportation logistics outsource helping you find capacity when you have a tough load to move be it high volume low price or high price low volume, we are your perfect provider, in managing B2B and B2C. Our asset-neutral approach allows us to design a transportation logistics solution that leverages multiple modes to ensure your product arrives where it needs to be at the lowest landed cost. We can put our carrier sourcing expertise to work in securing and managing the optimal combination of service providers or, we can manage your network of carriers.

With our core transportation solution, you'll also get a leading-edge Transportation Management System (TMS)/GPS configured and deployed to your specifications. We add strong operating procedures and seasoned personnel to help transform your transportation network into a competitive advantage.

Transportation Management Services:

(1) Network design and optimization;

(2) Carrier sourcing and compliance management;

(3) Transportation planning and optimization;

(4) Order consolidation services;

(5) Dedicated transportation;

(6) Freight bill audit and payment;

(7) Claims administration;

(8) Visibility and shipment event tracking;

(9) Performance management and reporting;

(10) Transportation management system (TMS) configuration and implementation.

8. 3. 5. 3 Supply Chain Management

Supply chain management is the management of the flow of goods and services and includes all processes that transform raw materials into final products. It involves the active streamlining of a business's supply-side activities to maximize customer value and gain a competitive advantage in the marketplace.

SCM represents an effort by suppliers to develop and implement supply chains that are as efficient and economical as possible. Supply chains cover everything from production to product development to the information systems needed to direct these undertakings.

Typically, SCM attempts to centrally control or link the production, shipment, and distribution of a product. By managing the supply chain, companies are able to cut excess costs and deliver products to the consumer faster. This is done by keeping tighter control of internal inventories, internal production, distribution, sales, and the inventories of company vendors.

SCM is based on the idea that nearly every product that comes to market results from the efforts of various organizations that make up a supply chain. Although supply chains have existed for ages, most companies have only recently paid attention to them as a value-add to their operations.

In SCM, the supply chain manager coordinates the logistics of all aspects of the supply chain which consists of five parts:

(1) Planning. Plan and manage all resources required to meet customer demand for a company's product or service. When the supply chain is established, determine metrics to measure whether the supply chain is efficient, effective, delivers value to customers and meets company goals.

(2) Sourcing. Choose suppliers to provide the goods and services needed to create the product. Then establish processes to monitor and manage supplier relationships. Key processes include ordering, receiving, managing inventory, and authorizing supplier payments.

(3) Manufacturing. Organize the activities required to accept raw materials, manufacture the product, test for quality, package for shipping and schedule for delivery.

(4) Delivery and logistics. Coordinate customer orders, schedule deliveries, dispatch loads, invoice customers and receive payments.

(5) Returning. Create a network or process to take back defective, excess, or unwanted products.

Effective supply chain management systems minimize cost, waste, and time in the production cycle. The industry standard has become a just - in - time supply chain where retail sales automatically signal replenishment orders to manufacturers. Retail shelves can then be restocked almost as quickly as product is sold. One way to further improve on this process is to analyze the data from supply chain partners to see where further improvements can be made.

By analyzing partner data, the CIO. com post identifies three scenarios where effective supply chain management increases value to the supply chain cycle:

(1) Identifying potential problems. When a customer orders more product than the manufacturer can deliver, the buyer can complain of poor service. Through data analysis, manufacturers may be able to anticipate the shortage before the buyer is disappointed.

(2) Optimizing price dynamically. Seasonal products have a limited shelf life. At the end of the season, these products are typically scrapped or sold at deep discounts. Airlines, hotels, and others with perishable "products" typically adjust prices dynamically to meet demand. By using analytic software, similar forecasting techniques can improve margins, even for hard goods.

(3) Improving the allocation of "available to promise" inventory. Analytical software tools help to dynamically allocate resources and schedule work based on the sales forecast, actual orders and promised delivery of raw materials. Manufacturers can confirm a product delivery date when the order is placed-significantly reducing incorrectly-filled orders.

词汇

生词	音标	释义
storage	[ˈstɔːrɪdʒ]	*n.* 贮存，贮藏（空间）；存储（方式）；付费托管
inventory	[ˈɪnvəntrɪ]	*n.* 清单；财产清单；存货，库存 *v.* 开列清单
procurement	[prəˈkjʊrmənt]	*n.*（尤指为政府或机构）采购，购买
disposal	[dɪˈspəuzl]	*n.* 去掉；清除；处理；（企业、财产等的）变卖，让与
algorithmic	[ˌælgəˈrɪðmɪk]	*adj.* 算法的，规则系统的
derive	[dɪˈraɪv]	*v.* 获得；取得；得到；（使）起源；（使）产生
antiquity	[ænˈtɪkwətɪ]	*n.* 古代（尤指古希腊和古罗马期）；古老；古；文物；古董；古迹
foodstuff	[ˈfuːdstʌf]	*n.* 食物；食品
maintenance	[ˈmeɪntənəns]	*n.* 维护；保养；维持；保持；（依法 应负担的）生活费；抚养费
methodology	[ˌmeθəˈdɒlədʒi]	*n.*（从事某一活动的）方法，原则

integrate [ˈɪntɪgreɪt] v. （使）合并，成为一体；（使）加 入，融入群体

insufficiency [ˌɪnsəˈfɪʃənsɪ] n. 不足，不充分，缺乏

simultaneously [ˌsɪməlˈteɪnɪəslɪ] adv. 同时；联立；急切地

constitute [ˈkɒnstɪtuːt] v. （被认为或看作）是；组成；（合法或正式地）成立，设立

conceiving [kənˈsiːvɪŋ] v. 想出（主意、计划等）；想象；构想；设想；怀孕；怀（胎）

execution [ˌeksɪˈkjuːʃn] n. 处决；实行；执行；实施；表演；（乐曲的）演奏；（艺术品的）制作

implementation [ɪmplɪmɛnˈteɪʃən] n. 执行，履行；实施，贯彻；生效；工具；仪器

mastery [ˈmæstərɪ] n. 精通；熟练掌握；控制；驾驭；控制力量

adhering [ədˈhɪrɪŋ] v. 黏附；附着

entity [ˈentəti] n. 独立存在物；实体

retrieve [rɪˈtriːv] v. 取回；索回；检索数据；扭转颓势；挽回；找回

commodity [kəˈmɒdətɪ] n. 商品；有用的东西；有使用价值的事物

echelon [ˈeʃəlɒn] n. 职权的等级；阶层；（士兵、飞机等的）梯形编队，梯队

dimension [daɪˈmenʃn] n. 维（构成空间的因素）；尺寸；规模；程度；范围；方面；侧面

deterministic [dɪˌtɜːrmɪˈnɪstɪk] adj. （思想）基于决定论的；（力量、因素）不可抗拒的

stochastic [stəˈkæstɪk] adj. 随机的；机会的；有可能性的；随便的

semiautomatic [ˌsɛmɪˌɔːtəˈmætɪk] adj. 半自动的；半自动手枪；半自动模式

refinery [rɪˈfaɪnərɪ] n. 炼油厂；制糖厂；精制厂

petrochemical [ˌpetrəʊˈkemɪkl] n. 石油化学产品

pharmaceutical [ˌfɑːməˈsuːtɪkl] adj. 制药的；配药的；卖药
n. 药物

implementing [ˈɪmplɪmentɪŋ] v. 使生效；贯彻；执行；实施

vaccine [vækˈsiːn] n. 疫苗；菌苗

serum [ˈsɪrəm] n. 血清；免疫血清；浆液（体液的水样部分）

immunization [ˌɪmjunaɪˈzeɪʃn] n. 免疫，免疫作用，免疫法

warranty [ˈwɒrəntɪ] n. （商品）保用单

refurbishment [ˌriːˈfɜːrbɪʃmənt] n. 翻新；整修

hybrid [ˈhaɪbrɪd] n. 杂种动物；杂交植物；杂种；混合物，合成物
adj. 混合的；杂种的

ultra-fast	[ˈʌltrə fæst]	*adj.* 超快的；超速的
hyperloop	[haɪpəluːp]	*n.* 超级高铁；超回路列车；超回路；超级回路
aerospace	[ˈeərəʊspeɪs]	*n.* 航空航天（工业）；航空航天技术
minimizing	[ˈmɪnɪmaɪzɪŋ]	*v.* 使减少到最低限度；降低；贬低；使显得不重要
synonym	[ˈsɪnəˌnɪmz]	*n.* 同义词
pallet	[ˈpælət]	*n.* 托盘；平台；运货板；（睡觉用的）草垫子
bulk	[bʌlk]	*n.* 主体；大部分；（大）体积；大（量）；巨大的体重（或重量、形状、身体等） *v.* 变得巨大（要）；胀大；增大；扩展；（重要性等）增加；用眼力估计；毛估
conveyor	[kənˈveɪə]	*n.* 运送者；传送者；传播者；传达者
tailor	[ˈteɪlə]	*adj.* 定做的；合身的；特制的；专门的 *v.* 专门制作；定做
omnipresent	[ˌɒmnɪˈpreznt]	*adj.* 无所不在的；遍及各处的
threshold	[ˈθreʃhoʊld]	*n.* 门槛；门口；阈；界；起始点；开端；起点；入门
thence	[ðens]	*adv.* 从那里；然后；随之
Gartner	[ˈgɑrtnər]	*n.* 高德纳公司；加特纳；加特讷；格特纳
transactional	[trænˈzækʃən(ə)l]	*adj.* 交易的，业务的；（社会交往中）相互作用的
warehouse	[ˈweəhaʊs]	*n.* 仓库；货栈；货仓
replenishment	[rɪˈplenɪʃmənt]	*n.* 补充；充满
prioritize	[praɪˈɔːrətaɪz]	*v.* 按重要性排列；划分优先顺序；优先处理
vendors	[ˈvɛndɔːz]	*n.* 小贩；摊贩；（某种产品的）销售公司；（房屋等的）卖主
consolidation	[kənˌsɒlɪˈdeɪʃn]	*n.* 合并；巩固，加强，联合；变坚固，变结实；充实
freight	[freɪt]	*n.*（海运、空运或陆运的）货物；货运 *v.* 寄送，运送（货物）；货运
audit	[ˈɔːdɪt]	*n.* 审计；稽核；（质量或标准的）审查，检查 *v.* 审计；稽核
configuration	[kənˌfɪgəˈreɪʃn]	*n.* 布局；结构；构造；格局；形状；（计算机的）配置
optimize	[ˈɒptɪmaɪz]	*v.* 使最优化；充分利用

长难句

Logistics is the time related positioning of resource or the strategic management of the total supply chain. The supply chain is a sequence of events intended to satisfy a customer. It can include procurement, manufacture, distribution, and waste disposal, together with associated transport, storage, and information technology.

物流是与时间相关的资源定位或整个供应链的战略管理。供应链是一系列旨在让顾客满意的事件。它可以包括采购、制造、分销和废物处理，以及相关的运输、储存和信息技术。

A Logistics Flow (LF) corresponds to any matter able to circulate throughout the components of a Logistics System (LS). Flow circulation can be made bottom-up to top-down or the other way round.

物流（LF）对应于任何能够在物流系统（LS）组件中循环的物质。物流循环可以自下而上到自上而下或相反。

Warehouse inventory management is the process by which stock stored in a warehouse or other storage facility is received, tracked, audited, and managed for order fulfilment.

仓库库存管理是指接收、跟踪、审计和管理仓库或其他存储设施中的库存以满足订单的过程。

Logistics management is the governance of supply chain management functions that helps organizations plan, manage, and implement processes to move and store goods.

物流管理是对供应链管理功能的综合治理，它帮助组织计划、管理以及实施移动和存储货物的过程。

Supply chain management is the management of the flow of goods and services and includes all processes that transform raw materials into final products.

供应链管理是对商品和服务流动的管理，包括将原材料转化为最终产品的所有过程。

Reference

[1] 景平．物流英语［M］．上海：复旦大学出版社，2010.
[2] 周跃进，任秉银．工业工程专业英语［M］．北京：机械工业出版社，2017.

9 Ergonomics

9.1 Introduction to Ergonomics

9.1.1 Definition of Ergonomics

Ergonomics (or human factors) is the scientific discipline concerned with the understanding of interactions among humans and other elements of a system, and the profession that applies theory, principles, data, and methods to design in order to optimize human well-being and overall system performance. It seeks to harmonize the functionality of tasks with the human requirements of those performing them. Ergonomic design focuses on the compatibility of objects and environments with the humans using them. The principles of ergonomic design can be applied to everyday objects and workspaces. The word "ergonomic" means human engineering. Ergonomic design is said to be human-centered design focusing on usability. It seeks to ensure that human restrictions and capabilities are met and supported by design options. In an ergonomic environment, equipment and tasks will be aligned.

Ergonomics is the practice of designing products, services, interfaces, and environments to suit the physical and cognitive characteristics of humans. The field relies on sciences such as psychology, biomechanics, physiology, and anthropometry. It has its roots in efforts to improve organizational performance and health & safety initiatives.

9.1.2 Origin and Development of Ergonomics

The term ergonomics originally comes from the Greek words ergon (work or labor) and nomos (natural laws). The fact that the word ergonomics was coined by a Polish scholar, Wojciech Jastrzębowski, in 1857 became widely known when his book in Polish was reprinted with English translation in 1997.

In Britain, the Ergonomic Society was formed in 1952 with people from psychology, biology, physiology, and design. In the United States, the Human Factors Society was formed in 1957. In the US "human factors engineering" was emphasized by the US military with concentration human engineering and engineering psychology. US efforts also focused the "role" of an individual within a complex system.

In fact, basic ergonomics has existed since the first ancestors of modern man began creating primitive tools to make tasks easier. Archaeological evidence from as far back as some of the earliest Egyptian dynasties, and other, more concrete findings from 5th Century BCE Greece, have shown that tools, household equipment, and other manmade devices illustrated sophisticated (for their time) ergonomic principles. Shortly after the Industrial Revolution, factory machinery and equipment started being built with design considerations closer to what we think of today as "ergonomics" . Most of those designs, however, were created to increase the speed and efficiency of production, rather than to create comfort and/or ease of use for the workers involved Ergonomics in the modern sense began to become more widespread during World War Ⅱ. Military equipment, machinery, and weaponry specifically airplanes -was becoming increasingly complex. After the innovations of World War Ⅱ, ergonomics continued to flourish, as its principles were further applied to evolving technologies.

The science of modern ergonomics includes the work of industrial engineers, occupational medical physicians, safety engineers, and many others studying both "cognitive ergonomics" (human behavior, decision making processes, perception relative to design, etc.) and "industrial ergonomics" (physical aspects of the workplace, human physical abilities, etc.). Nearly every aspect of modern life now includes some level of ergonomic design. Automobile interiors, kitchen appliances, office chairs and desks, and other frequently used devices are designed ergonomically. Even the machines and tools used to build and assemble those devices are superbly ergonomic. By maximizing efficiency and, more importantly, user comfort and safety, ergonomics continues to make life easier.

9.1.3 Disciplines Components of Ergonomics

The science of ergonomics promotes a holistic approach which considers the physical, cognitive, and organizational environment. Each of these components of ergonomics has a specific set of considerations. Figure 9-1 shows concept of ergonomics. Ergonomics draws on a number of scientific disciplines, including:

(1) Anthropometry;

(2) Biomechanics;

(3) Mechanical engineering;

(4) Industrial engineering;

(5) Industrial design;

(6) Information design;

(7) Kinesiology;

(8) Physiology;

(9) Psychology[7].

These disciplines contribute to the design and evaluation of tasks, products, environments, and systems in order to make them compatible with the needs, abilities, and limitations of humans.

9.1.4 Research Contents of Ergonomics

Ergonomics includes study of the following:

(1) Work environment.

1) Physical demands (e.g., lifting objects, moving objects);

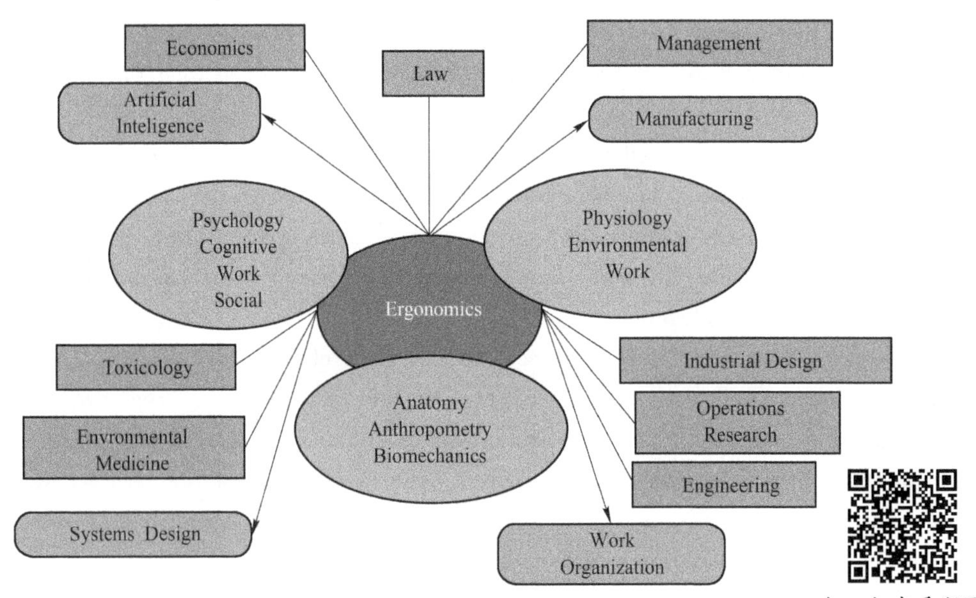

Figure 9-1 Concept of ergonomics

2) Skill demands (e. g. , typing at 110 words per minute);

3) Risk demands (e. g. , running on an ice pavement);

4) Time demands (e. g. , trying to finish all of the work by the end of semester).

(2) Psychosocial environment.

1) Social (e. g. , working in teams);

2) Cultural (e. g. , pace of life is different in different countries);

3) Lifestyle (e. g. , work vs. leisure time, and quality of life issues differ between countries).

(3) Physical environment.

1) Physical agents (e. g. , heat, noise, vibration);

2) Chemical agents (e. g. , air pollutants);

3) Biological agents (e. g. , airborne diseases).

(4) Technology.

1) Product design (e. g. , designing product dimensions using anthropometrics, biomechanics data);

2) Hardware Interface design (e. g. , designing controls and displays to meet user expectations);

3) Software interface design (e. g. , designing icons and commands to meet user expectations).

9.1.5 Ergonomics Domains of Specialization

According to the international ergonomics association, there are three broad domains of ergonomics: physical, cognitive, and organizational.

(1) Physical ergonomics.

Physical ergonomics considers human anatomical, anthropometric, physiological, and

biomechanical characteristics as they relate to physical activity. The consequences of repetitive motion, vibration, force, working postures and the environment are the most common areas of consideration for physical ergonomics. **Other factors** are shown in Figure 9-2.

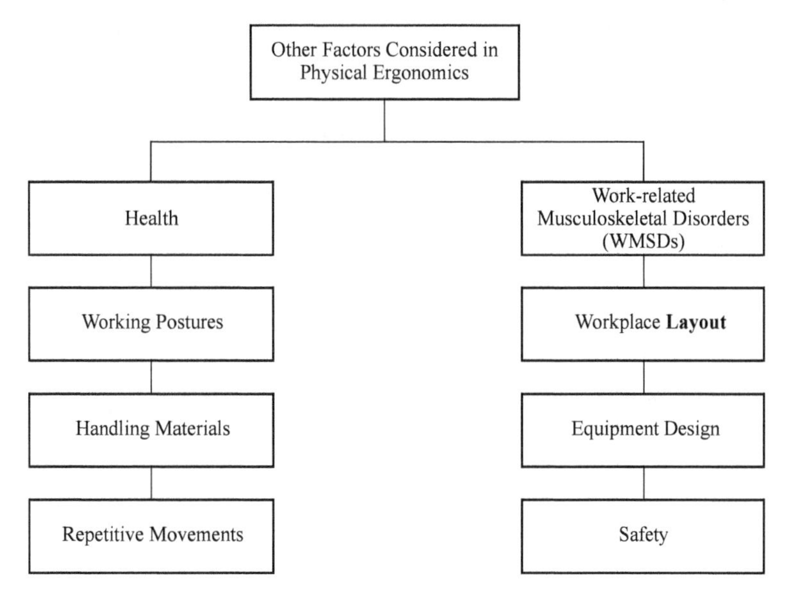

Figure 9-2 Other factors considered in physical ergonomics

(2) Cognitive ergonomics.

Cognitive ergonomics considers human cognitive abilities and limitations while working. Mental processes, such as perception, attention, memory, reasoning, decision-making, learning and motor response, are considered as they affect interactions among humans and other mechanical elements of a system. Cognitive ergonomics are shown in Figure 9-3.

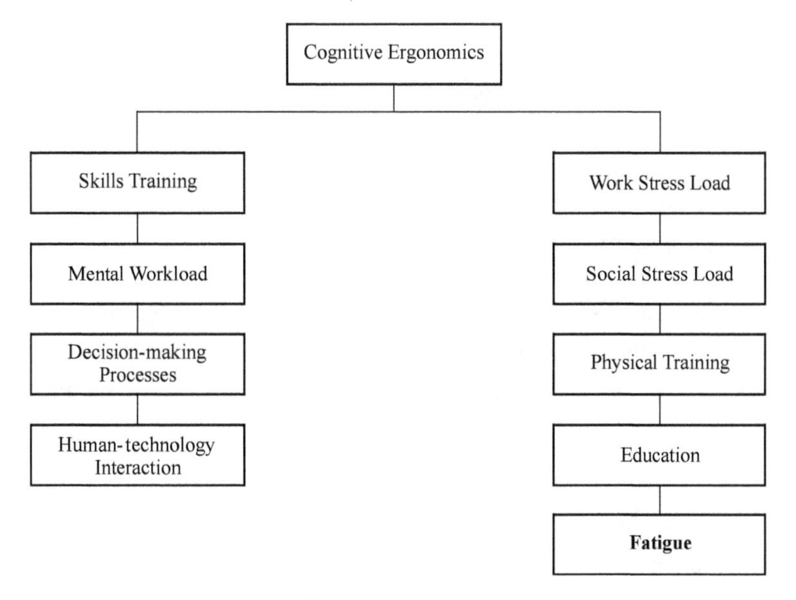

Figure 9-3 The content of cognitive ergonomics

(3) Organizational ergonomics.

Organizational ergonomics considers the structures, policies, and processes of any organization. **The goal of organizational ergonomics is to achieve a harmonized system, taking into consideration the consequences of technology on human relationships, processes, and organizations.** Examples of organizational ergonomics are shown in Figure 9-4.

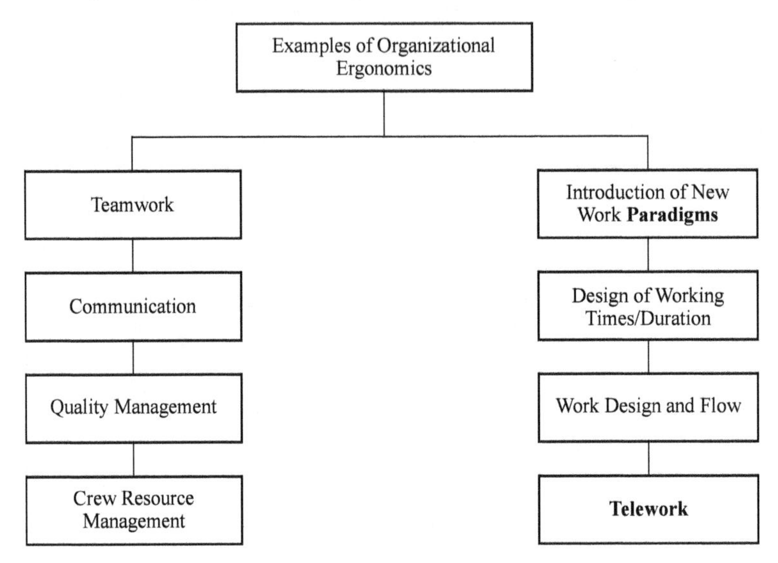

Figure 9-4 Examples of organizational ergonomics

Each aspect of physical, cognitive, and organizational ergonomics can be applied individually, or more successfully in conjunction with one another. While these lists may appear **daunting** in their entirety, rest assured: addressing even one area will prove beneficial in reducing injury rates.

9.2 Foundation of Ergonomics

9.2.1 Anthropometry

9.2.1.1 Anthropometry Definition

Anthropometry is the science of obtaining systematic measurements of the human body. Anthropometry first developed in the 19th century as a method employed by physical anthropologists for the study of human variation and evolution in both living and extinct populations. In particular, such anthropometric measurements have been used historically as a means to associate racial, cultural, and psychological attributes with physical properties. Specifically, anthropomorphic measurements involve the size (e.g., height, weight, surface area, and volume), structure (e.g., sitting vs. standing height, shoulder and hip width, arm/leg length, and neck circumference), and composition (e.g., percentage of body fat, water content, and lean body mass) of humans. Figure 9-5 shows basic anthropometry.

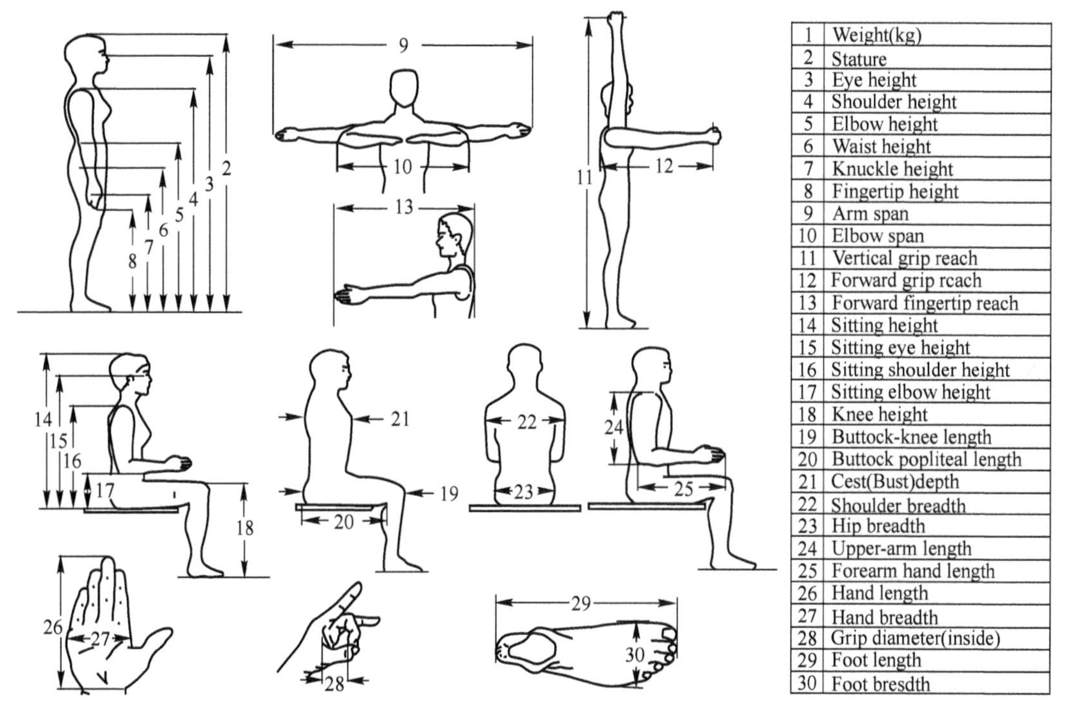

Figure 9-5 Basic anthropometry

9.2.1.2 Anthropometric Tools

To obtain anthropometric measurements, a variety of specialized tools (as depicted below) are used:

(1) Stadiometers: height;

(2) Anthropometers: length and circumference of body segments;

(3) Bicondylar calipers: bone diameter;

(4) Skinfold calipers: skin thickness and subcutaneous fat;

(5) Scales: weight.

Although the majority of the instruments appear straight forward to use, a high level of training is required to achieve high validity and accuracy of measurements. Figure 9-6 shows anthropometry tools.

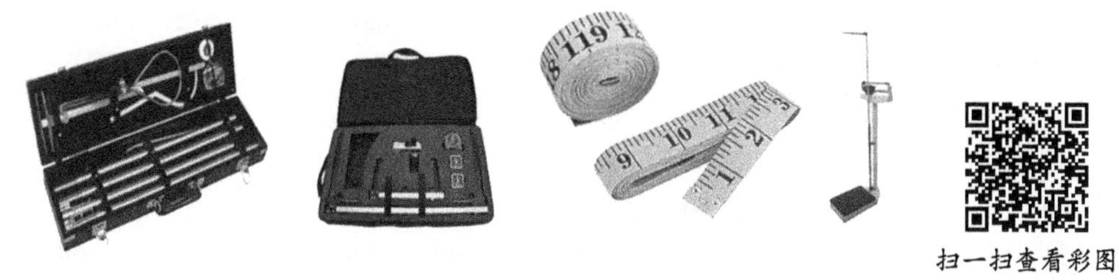

Figure 9-6 Anthropometry tools

9.2.1.3 Current Applications of Anthropometry

While physical anthropologists and criminologists continue to use anthropometric measurements in the study of human evolution through the comparison of novel fossil remains to archived specimens and forensics, respectively, current applications have extended to:

(1) Industrial design and architecture (e.g., vehicle seating and cockpits).

(2) Clothing (e.g., military uniforms).

(3) Ergonomics (e.g., seating).

(4) Medicine (e.g., nutrition, aging, obesity, sports science, and diabetes).

In these industries, anthropometric data is invaluable to the optimization of various products and observing the changes which occur in response to various lifestyle, genetic, and ethic factors.

9.2.2 Physiology and Psychology

9.2.2.1 Physiology

Physiology is the study of how the human body works. It describes the chemistry and physics behind basic body functions, from how molecules behave in cells to how systems of organs work together. It helps us understand what happens in a healthy body in everyday life and what goes wrong when someone gets sick. Most of physiology depends on basic research studies carried out in a laboratory. Figure 9-7 shows six systems of human body.

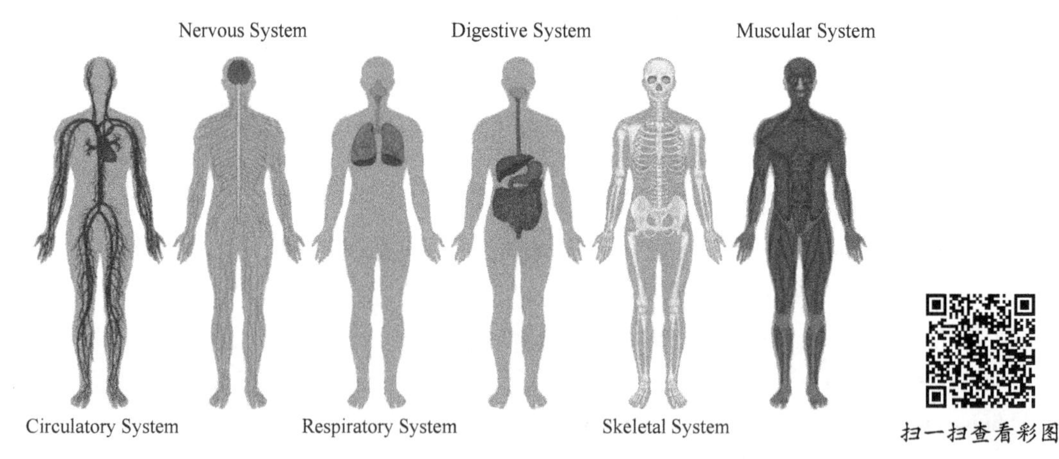

Figure 9-7 Six systems of human body

Some physiologists study single proteins or cells, while others might do research on how cells interact to form tissues, organs, and systems within the body.

Doctors use physiology to learn more about many different organ systems, including:

(1) The cardiovascular system (your heart and blood vessels);

(2) The digestive system (the stomach, intestines, and other organs that digest food);

(3) The endocrine system (glands that make hormones, the chemicals that control many body functions);

(4) The immune system (your body's defense against germs and disease);

(5) The muscular system (the muscles you use to move your body);

(6) The nervous system (your brain, spinal cord, and nerves);

(7) The renal system (your kidneys and other organs that control the fluid in your body);

(8) The reproductive system (sex organs for men and women);

(9) The respiratory system (your lungs and airways);

(10) The skeletal system—bones, joints, cartilage, and connective tissue.

For each system, physiology sheds light on the chemistry and physics of the structures involved. For example, physiologists have studied the electrical activity of cells in the heart that control its beat. They're also exploring the process by which eyes detect light, from how the cells in the retina process light particles called photons to how the eyes send signals about images to the brain.

9.2.2.2 Psychology

Psychology is the scientific study of mind and behavior. It is a multifaceted discipline of the sciences and includes many sub-fields of study, such as human development, social behavior, and cognitive processes. Psychology focuses on understanding a person's emotions, personality, and mind through scientific studies, experiments, observation, and research. The study of psychology has four goals: describe, explain, predict, and change/control.

(1) Describe.

We describe things every day with no conscious thought or effort, but the describing of psychology has a slightly different meaning than the describing we do in our day-to-day lives. Accurately describing a problem, an issue, or a behavior is the first goal of psychology. Descriptions help psychologists to distinguish between typical and atypical behaviors and gain more accurate understandings of human and animal behaviors and thoughts. A variety of research methods, including case studies, surveys, self-tests, and natural observation allow psychologists to pursue this goal.

(2) Explain.

In addition to describing, psychologists seek to be able to explain behaviors. The goal of explaining is to provide answers to questions about why people react to certain stimuli in certain ways, how various factors impact personalities and mental health, and so on. Psychologists often use experiments, which measure the impacts of variables upon behaviors, to help formulate theories that explain aspects of human and animal behaviors.

(3) Predict.

Making predictions about how humans and animals will think, and act is the third goal of psychology. By looking at past observed behavior (describing and explaining), psychologists aim to predict how that behavior may appear again in the future, as well as whether others might exhibit the same behavior.

(4) Change/Control.

Psychology aims to change, influence, or control behavior to make positive, constructive, meaningful, and lasting changes in people's lives and to influence their behavior for the better. This is the final and most important goal of psychology. These four goals of psychology are not that different from how you would naturally interact with others. Suppose, for example, that someone did something they weren't supposed to do, and this action had a negative impact on their life.

9.2.3 Work Environment

(1) The relationship between man and environment.

Man and environment are inter-related. The environment influences the life of human beings and also human beings modify their environment as a result of their growth, dispersal, activities, death, and decay etc. Thus all living beings including man and their environment are mutually reactive affecting each other in a number of ways and a dynamic **equilibrium** is possible in between the two, i. e. human beings (society) and environment are interdependent. Figure 9-8 shows work environment.

Figure 9-8 Work environment

(2) Impacts of working environment.

The research work in the area of ergonomics consists of impacts of environment-lighting, noise, vibration and so on. The interactions of these stimulates are factored in the developmental, operational test & evaluations and product design.

Vision is usually the primary channel for information, yet systems are often so poorly designed that the user is unable to see the work area clearly. Many workers using computers cannot see their screens because of glare or reflections. Others, doing precise assembly tasks, have insufficient lighting and suffer eyestrain and reduced output as a result.

Sound can be a useful way to provide information, especially for warning signals. However, care must be taken not to overload this sensory channel. A recent airliner had 16 different audio warnings, far too many for a pilot to deal with in an emergency situation. A more sensible approach was to have just a few audio signals to alert the pilot to get information guidance from a visual display.

(3) Environmental factors.

It is essential that as an employee you receive a suitable amount of natural light when sitting at your desk and when anywhere in your office. You must also be able to control the amount of natural light in the office with the use of blinds or shades. Apart from natural light, electric light can have a major effect. Strong overhead lighting that can cause headaches, eyestrain and fatigue can be reduced by simply adding filters or introducing lower in direct lighting. Eyestrain and headaches can appear as a result of the glare you get on your monitor screen. Implementing such lighting and shading will enable you to reduce the uneasiness. A lighting scheme suitably designed for a working environment can enhance the safety and operational effectiveness of our servicemen and women. Lighting specialists support infrastructure and platform project by evaluating and designing lighting schemes in man-made or artificial environments based on man's psychological reactions to color and light.

9.2.4 Man-Machine System

9.2.4.1 Characteristics of Man-Machine System

Characteristics of man-machine system are as follows:

(1) The man-machine system consists of the man, the machine and system environment.

(2) It is essentially artificial by nature and is specifically developed to fulfill some purpose or specific aim.

(3) It has specific inputs and outputs which are appropriately balanced.

(4) It is variable in size and complexity and is dynamic in performance.

(5) Subsystems of man machine system interact with and effects the other parts.

(6) The man-machine system becomes more efficient when inputs and out puts are adequately balanced.

(7) Environmental factors or system environment effects system performance.

9.2.4.2 Classification of Man-Machine Systems

Depending upon size and complexity, man-machine systems are of following three types:

(1) Manual systems:

They are essentially man directed systems. These are flexible in nature and small in size. Simple tools and equipment are used, and the efficiency is dependent upon the human factor. A large variability is possible in a manual system as every worker may select different method to do the same job.

(2) Mechanical systems:

They are more complex and inflexible in nature than manual systems. The machine component is power driven and human activity is information processing, decision making and controlling occasionally knows semi-automatic systems, they have components which are well integrated. This is the feature which renders these systems rather inflexible. An automobile and a machine tool operated by driver or operator are good examples of his class.

(3) Automatic systems:

A complex system in which all operational functions are performed by automatic devices is known as automatic system. Operational functions are sensing information processing decision making and action. It is completely inflexible in nature and cannot be adopted to uses other that the one for which it has be designed.

The human element/component performs the jobs of monitoring, programming the function, maintenance, and upkeep. An automatic telephone exchange, a digital computer and automatic screw cutting, machines are good examples of automatic systems. A perfectly reliable automatic system does not exist at present.

9.2.5 An Ergonomic Case Study for Workers at Siemens Automotive

(1) Task prior to abatement (description):

Siemens auburn hills, mich-based, designs and manufactures electronic automobile systems. It has 14,000 employees. Workers had to lean to the right and grope for the mouse on the desktop which was away from the keyboard. Workers had to look sharply down and to the left to read documents on the worktable while using the typewriter. Although Siemens purchased adjustable workstations, they were not easily or completely adjustable.

(2) Task prior to abatement (method which identified hazard):

1) Complaining from 43 of 100 employees of pain in their shoulders, back, ebow and fingers.

2) Increasing incidence rate of carpal tunnel syndrome.

3) Scheduling 3 of 43 employees for cervical laminectomies.

4) Several employees visited chiropractors.

(3) Ergonomic risk factor (posture):

1) Workers had to lean to the right and grope for the mouse on the desktop and away from the keyboard.

2) Workers had to stretch their necks to read the documents.

(4) Ergonomic solution (administrative controls):

1) Performing ergonomic evaluations on workstations of those employees with symptoms.

2) Training for stretch exercises.

3) Encouraging the CAD operators to take frequent, short breaks.

4) Conducting a back school training program twice a year to emphasize good lifting and pushing techniques, good posture and exercises for strength and flexibility.

5) Developing a slide show on office ergonomics for new employees which includes training on how to adjust their workstations.

(5) Ergonomic solution (engineering controls):

1) Encouraging the CAD operators to wear 22-inch focus glasses.

2) Purchasing 27 back cushions, 71 lumbar supports in three different sizes, 24 keyboard/mouse rests and 12 document holders in the past five years.

3) Providing adjustable chairs.

4) Providing a footrest and raising the chair for shorter workers.

(6) Ergonomic solution (benefits):

1) All workers that perform the tasks now have reduced exposure to CTD risk factors.

2) Saving 20,000 hours lost time per year since eliminating CTD-related complaints in two years.

(7) Ergonomic solution (cost):

The total cost of ergonomics interventions was about $3,600.

(8) Ergonomic solution (method which verified effectiveness):

Eliminating CTD-related complaints in two years.

词汇

生词	音标	释义
harmonize	[ˈhɑːmənaɪz]	*v.* 协调，和谐；使相一致；使协调
align	[əˈlaɪn]	*v.* 排整齐；校准；（尤指）使成一条直线；使一致
biomechanics	[ˌbaɪəʊməˈkænɪks]	*n.* 生物力学
anthropometry	[ˌænθrəˈpɒmɪtrɪ]	*n.* 人体测量，人体测量学
initiative	[ɪˈnɪʃətɪv]	*n.* 倡议；新方案；主动性；积极性；自发性；主动权
primitive	[ˈprɪmətɪv]	*adj.* 原始的；远古的；落后的
archaeological	[ˌɑːkɪəˈlɒdʒɪkəl]	*adj.* 考古的；考古学的；考古学上的
Egyptian	[ɪˈdʒɪpʃn]	*n.* 埃及人；古埃及人 adj. 埃及的；埃及人的
sophisticate	[səˈfɪstɪkeɪt]	*n.* 老于世故的人；见多识广的人
widespread	[ˈwaɪdspred]	*adj.* 分布广的；普遍的；广泛的
weaponry	[ˈwepənrɪ]	*n.* 武器；兵器
flourish	[ˈflʌrɪʃ]	*v.* 繁荣；昌盛；兴旺
hypothetical	[ˌhaɪpəˈθetɪkl]	*adj.* 假设的；假定的
cognitive	[ˈkɒgnətɪv]	*adj.* 认知的；感知的；认识的
interior	[ɪnˈtɪərɪə(r)]	*n.* 内部；里面；内陆 *adj.* 内部的；里面的
holistic	[həˈlɪstɪk]	*adj.* 整体的；全面的；功能整体性的
kinesiology	[ˌkɪnɪˌsɪˈɑlədʒɪ]	*n.* 运动疗法；运动学；人体运动学
evaluation	[ɪˌvæljuˈeɪʃn]	*n.* 定值；估计；评价；评审
compatible	[kəmˈpætəbl]	*adj.* 可共用的；兼容的；可共存的
pavement	[ˈpeɪvmənt]	*n.* 人行道；石板铺的地面；路面
vibration	[vaɪˈbreɪʃn]	*n.* 震动；颤动；抖动
airborne	[ˈeəbɔːn]	*adj.* 升空；空气传播的；空降的
repetitive	[rɪˈpetətɪv]	*adj.* 重复乏味的；多次重复的
layout	[ˈleɪaʊt]	*n.* 布局；布置；设计；安排
cognitive	[ˈkɒgnətɪv]	*adj.* 认知的；感知的；认识的

fatigue	[fə'tiːg]	n. 疲劳；劳累；厌倦 v. 使疲乏；使劳累
paradigm	['pærədaɪm]	n. 典范；范例；样式；词形变化表
daunt	[dɔːnt]	v. 使胆怯；使气馁；使失去信心
stadiometer	[steɪdɪ'ɒmɪtə]	n. 测距仪；曲线长度仪器；自记经纬仪
anthropometer	[ˌænθrəpə'miːtə]	n. 人体测量器；人体测量仪
bicondylar		n. 双髁
caliper	['kæləpə]	n. 卡尺；卡钳 v. 用测径规测量；用卡钳测量
skinfold		n. 皮褶厚度；皮肤褶；皮脂厚度；皮褶
validity	[və'lɪdətɪ]	n. 有效，合法性；认可
criminologist	[ˌkrɪmɪ'nɒlədʒɪst]	n. 刑事学家；犯罪学家
fossil	['fɒsl]	n. 化石；老人；（尤指）老顽固
cardiovascular	[ˌkɑːdɪəʊ'væskjələ（r）]	adj. 心血管的
digestive	[daɪ'dʒestɪv]	adj. 消化的；和消化有关的 n. 消化剂
endocrine	['endəʊkrɪn]	adj. 内分泌腺的；内分泌的
immune	[ɪ'mjuːn]	adj. 有免疫力；不受影响；受保护；免除；豁免
muscular	['mʌskjələ（r）]	adj. 肌肉的；强壮的；肌肉发达的
respiratory	[rə'spɪrətrɪ]	adj. 呼吸的
skeletal	['skelətl]	adj. 骨骼的；骨瘦如柴的
multifaceted	[ˌmʌltɪ'fæsɪtɪd]	adj. 多方面的；要从多方面考虑的
equilibrium	[ˌiːkwɪ'lɪbrɪəm]	n. 平衡；均衡；均势
decay	[dɪ'keɪ]	n. 腐烂；腐朽 v. （使）腐烂，腐朽
implement	['ɪmplɪment]	v. 使生效；贯彻；执行
classification	[ˌklæsɪfɪ'keɪʃn]	n. 分类；归类；分级；类别；等级
manual	['mænjuəl]	adj. 手工的；体力的 n. 使用手册；说明书

长难句

Ergonomics (or human factors) is the scientific discipline concerned with the understanding of interactions among humans and other elements of a system, and the profession that applies theory, principles, data, and methods to design in order to optimize human well-being and overall system performance.

人机工程学（或人因学）是一门科学学科，研究人与系统其他要素之间的相互作用，是应用理论、原理、数据和方法进行设计以优化人类幸福和系统整体性能的专业。

Ergonomics is the practice of designing products, services, interfaces, and environments to suit the physical and cognitive characteristics of humans.

人机工程学是设计产品、服务、界面和环境以适应人类的身体和认知特征的实践。

The science of ergonomics promotes a holistic approach which considers the physical, cognitive, and organizational environment. Each of these components of ergonomics has a specific set of considerations.

人机工程学的科学提倡一种综合考虑人体、认知和组织环境的方法，人机工程学的每一个组成部分都有一套特定的考虑因素。

Physical ergonomics considers human anatomical, anthropometric, physiological, and biomechanical characteristics as they relate to physical activity.

人体工效学考虑人体解剖学、人体测量学、生理学和生物力学特征，因为它们与身体活动有关。

The goal of organizational ergonomics is to achieve a harmonized system, taking into consideration the consequences of technology on human relationships, processes, and organizations.

组织工效学的目标是实现一个协调的系统，同时考虑到技术对人际关系、过程和组织的影响。

Physiology is the study of how the human body works. It describes the chemistry and physics behind basic body functions, from how molecules behave in cells to how systems of organs work together.

生理学是研究人体如何工作的学科。它描述了人体基本功能背后的化学和物理，从分子在细胞中的行为到器官系统如何协同工作。

10 Quality Management

10.1 Quality Management's Concept

10.1.1 Basic Concepts of Quality Management

Quality management is the means of implementing and carrying out quality policy. They perform goal planning and manage quality control and quality assurance activities. Quality management is responsible for seeing that all quality goals and objectives are implemented and that corrective actions have been achieved. They periodically review the quality system to ensure effectiveness and to identify and review any deficiencies. It is the act of overseeing all activities and tasks that must be accomplished to maintain a desired level of excellence. This includes the determination of a quality policy, creating and implementing quality planning and assurance, and quality control and quality improvement. It is also referred to as total quality management (TQM).

The key takeaways:

(1) Quality management is the act of overseeing all activities and tasks needed to maintain a desired level of excellence.

(2) Quality management includes the determination of a quality policy, creating and implementing quality planning and assurance, and quality control and quality improvement.

(3) TQM requires that all stakeholders in a business work together to improve processes, products, services, and the culture of the company itself.

The most famous example of TQM is Toyota's implementation of the Kanban system. A Kanban is a physical signal that creates a chain reaction, resulting in a specific action. Toyota used this idea to implement its just-in-time (JIT) inventory process. To make its assembly line more efficient, the company decided to keep just enough inventory on hand to fill customer orders as they were generated. Figure 10-1 shows quality inspection on assembly line.

Therefore, all parts of Toyota's assembly line are assigned a physical card that has an associated inventory number. Right before a part is installed in a car, the card is removed and moved up the supply chain, effectively requesting another of the same part. This allows the company to keep its inventory lean and not overstock unnecessary assets.

10.1.2 Development of Quality Management

In the 20th century, mankind entered the industrialization era characterized by "mechanization of processing, scale of operation, and monopoly of capital". In the past century, the development of

Figure 10-1 Quality inspection on assembly line

quality management has roughly gone through three stages:

(1) Quality inspection stage.

At the beginning of the 20th century, people's understanding of quality management was still limited to quality inspection. The methods used in quality inspection are various testing equipment and instruments, and the method is to strictly check and carry out 100% inspection. Meanwhile, the "scientific management movement" represented by Taylor appeared in the United States. "Scientific management" puts forward the requirement of scientific division of labor among personnel, and separates the planning function from the executive function, and adds an inspection link in the middle to supervise and inspect the implementation of the plan, design, product standards and other items. That is to say, planning design, production operation, inspection and supervision are each responsible for special personnel, which has created a full-time inspection team and constituted a full-time inspection department. In this way, the quality inspection agency is independent. At first, people emphasized the foreman's role in quality assurance, and transferred the responsibility of quality management from the operator to the foreman, so it was called "the foreman's quality management".

(2) Statistical quality control stage.

After the end of the Second World War, many American companies expanded their production scale. In addition to the original munitions factories that continued to implement quality management conditions, many civilian industries have also adopted this method. Many countries outside the United States, such as Canada and France, Germany, Italy, Mexico, and Japan have also successively implemented statistical quality management and achieved results. However, statistical quality management also has shortcomings. It overemphasizes the statistical methods of quality control, which makes people mistakenly believe that "quality management is statistical methods" and "quality management is the business of statistical experts". It makes most people feel unattainable and daunting. At the same time, its quality control and management are limited to manufacturing and inspection departments, ignoring the impact of other departments' work on quality. In this way, the enthusiasm of various departments and employees cannot be fully utilized, which restricts its promotion and application. The resolution of these problems pushes the quality management to a new stage.

(3) Total quality management stage.

Since the 1950s, productive forces have developed rapidly, science and technology have changed with each passing day, and many new situations have emerged. There are mainly the following aspects:

With the development of science and technology and industrial production, the requirements for quality are getting higher and higher. Since the 1950s, large, sophisticated, and complex products such as rockets, spacecraft, and artificial satellites have appeared. The requirements for product safety, reliability, and economy have become higher and higher, and the quality problems have become more prominent. People are required to use the concept of "system engineering" to comprehensively analyze and study quality issues as an organic whole, and implement management of all employees, entire processes, and entire enterprises.

In the 1960s, a "behavioral science theory" appeared in management theory, advocating to improve interpersonal relationships, mobilize people's enthusiasm, highlight the "value of human factors", and pay attention to the role of people in management.

With the intensification of market competition, especially in the international market, companies in various countries attach great importance to the issues of "product responsibility" and "quality assurance", strengthen internal quality management, and ensure the safe and reliable use of products.

10.1.3 W. Edwards Deming—The Father of Modern Quality Management

William Edwards Deming (October 14, 1900—December 20, 1993) was an American engineer, statistician, professor, author, lecturer, and management consultant. Educated initially as an electrical engineer and later specializing in mathematical physics, he helped develop the sampling techniques still used by the U. S. department of the census and the bureau of labor statistics.

In his book the new economics for industry, government, and education[1] Deming championed the work of walter shewhart, including statistical process control, operational definitions, and what Deming called the "shewhart cycle"[2], which had evolved into Plan-Do-Check-Act (PDCA). Deming is best known for his work in Japan after WW Ⅱ, particularly his work with the leaders of Japanese industry. That work began in July and August 1950, in Tokyo and at the Hakone convention center, when Deming delivered speeches on what he called "statistical product quality administration". Many in Japan credit Deming as one of the inspirations for what has become known as the Japanese post-war economic miracle of 1950 to 1960, when Japan rose from the ashes of war on the road to becoming the second-largest economy in the world through processes partially influenced by the ideas Deming taught[3]:

(1) Better design of products to improve service;

(2) Higher level of uniform product quality;

(3) Improvement of product testing in the workplace and in research centers;

(4) Greater sales through side [global] markets.

It is a common myth to credit Plan-Do-Check-Act (PDCA) to Deming. Deming referred to the PDCA cycle as a "**corruption**"[4]. Figure 10-2 shows PDCA deming worked from the shewhart cycle and over time eventually developed the Plan-Do-Study-Act (PDSA) cycle, which has the idea of **deductive** and **inductive** learning built into the learning and improvement cycle. Deming finally published the PDSA cycle in 1993, in the new economics on p. 132. Deming has added to the myth that he taught the Japanese the PDSA cycle with this quote on p. 247, " The PDSA cycle originated in my teaching in Japan in 1950. It appeared in the booklet Elementary Principles of the Statistical Control of Quality (JUSE, 1950: out of print)."

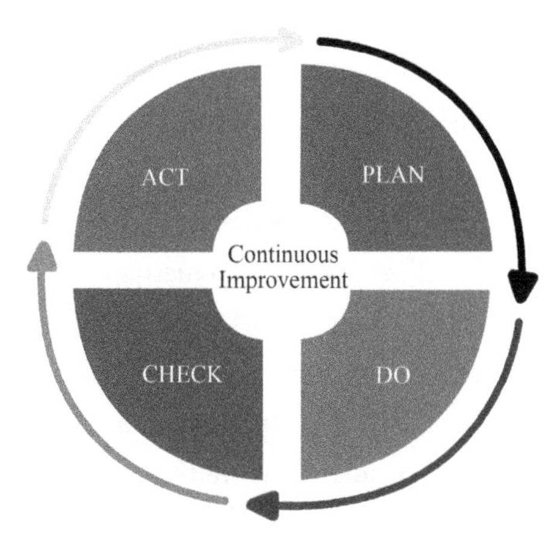

Figure 10-2 PDCA

扫一扫查看彩图

10.1.4 Joseph M. Juran—Quality Trilogy

Joseph Moses Juran (December 24, 1904 - February 28, 2008) was a Romanian-American engineer and management consultant. He was an evangelist for quality and quality management, having written several books on those subjects[5].

Juran was born in Brăila, Romania, one of the six children born to Jakob and Gitel Juran; they later lived in Gura Humorului. He had three sisters: Rebecca (nicknamed Betty), Minerva, who earned a doctoral degree and had a career in education, and Charlotte. He had two brothers: film and art director Nathan Juran, and Rudolph, known as Rudy. Rudy founded a municipal bond company[6]. In 1912, Joseph Juran emigrated to America with his family, settling in Minneapolis, Minnesota. He excelled in school, especially in mathematics. He was a chess champion at an early age[7], and dominated chess at Western Electric. Juran attended Minneapolis South High School where he graduated in 1920.

In 1924, with a bachelor's degree in electrical engineering from the University of Minnesota, Juran joined Western Electric's Hawthorne Works. His first job was troubleshooting in the Complaint Department. In 1925, Bell Labs proposed that Hawthorne Works personnel be trained in its newly developed statistical sampling and control chart techniques. Juran was chosen to join the Inspection Statistical Department, a small group of engineers charged with applying and disseminating Bell Labs' statistical quality control innovations. This highly visible position fueled Juran's rapid ascent in the organization and the course of his later career.

(1) Pareto principle.

In 1941, Juran stumbled across the work of Vilfredo Pareto and began to apply the Pareto principle to quality issues (for example, 80% of a problem is caused by 20% of the causes). This is also known as "the vital few and the trivial many" . In later years, Juran preferred "the vital few and the useful many" to signal that the remaining 80% of the causes should not be totally ignored.

For example, he argued that most defects are the result of a small percentage of the causes of all defects, according to the Economist. For another, 20% of a team's members are going to make up 80% of a project's successful results. And 20% of a businesses' customers will create 80% of the profit.

Juran felt organizations, armed with that knowledge, would focus less on meaningless minutiae and more on identifying the 20%. That means eliminating the 20% of mistakes causing the majority of defects, rewarding the 20% of employees causing 80% of the success and serving the 20% of loyal customers that drive sales. In a way, Pareto's Principle puts numbers to the idea that in business, as in life, things are not evenly distributed. Pareto was studying land ownership in Italy. But Juran saw that it applied to business, as well.

(2) The Juran trilogy.

Juran is one of the first to write about the cost of poor quality. This was illustrated by his "Juran trilogy," an approach to cross-functional management, which is composed of three managerial processes: quality planning, quality control, and quality improvement. Without change, there will be a constant waste; during change there will be increased costs, but after the improvement, margins will be higher, and the increased costs are recouped.

10.1.5 Philip Crosby—Father of Zero Defects

Philip Bayard "Phil" Crosby, (June 18, 1926—August 18, 2001) was a businessman and author who contributed to management theory and quality management practices[8].

Crosby initiated the zero defects program at the martin company. As the quality control manager of the pershing missile program, Crosby was credited with a 25 percent reduction in the overall rejection rate and a 30 percent reduction in scrap costs.

In 1979, Crosby started the management consulting company Philip Crosby Associates, Inc. This consulting group provided educational courses in quality management both at their headquarters in winter park, florida, and at eight foreign locations. Also, in 1979, Crosby published his first business book, quality is free. This book would become popular at the time because of the crisis in North American quality. During the late 1970s and into the 1980s, North American manufacturers were losing market share to Japanese products largely due to the superior quality of the Japanese goods.

Crosby's response to the quality crisis was the principle of "doing it right the first time" (DIRFT). He also included four major principles:

(1) The definition of quality is **conformance** to requirements (requirements meaning both the product and the customer's requirements).

(2) The system of quality is prevention.

(3) The performance standard is zero defects (relative to requirements).

(4) The measurement of quality is the price of nonconformance.

His belief was that an organization that establishes good quality management principles will see savings returns that more than pay for the cost of the quality system: "quality is free". It is less expensive to do it right the first time than to pay for rework and repairs.

10.1.6 Development of Quality Development in China

Since the reform and opening up, the Chinese economy has expanded much faster than the rest of the world, partly due to China's three-step development strategy aiming to quadruple the 1980 gross national output by the end of the 20th century. On its path to becoming a modernized country, China had also set targets to double its economic size in the first two decades of this century. The targets contributed to the brisk growth of past decades and helped China become the world's second largest economy. But in a report delivered to the 19th National Congress of the Communist Party of China, there was no such target. China has entered a "new era," and from here on things are different.

This shows that China will prioritize the quality of development rather than fast economic expansion.

The attitude is in line with the central leadership's judgment in 2014 that China's economy had entered a "new normal," featuring medium-high growth rather than fast growth, upgrading of economic structure and innovation.

China's economy grew at an average annual rate of 9.8 percent between 1979 and 2012, dwarfing the global growth of 2.8 percent in the same period.

However, the fast growth was accompanied by excessive use of resources, environmental pollution, and overcapacity.

Since 2012, the country has showed greater tolerance of lower growth rates in the pursuit of a better structure, quality, and efficiency. It lowered the annual GDP growth target from 7.5 percent for 2012—2014 to 6.5 percent this year.

Average annual growth between 2013 and 2016 slowed to 7.2 percent, still much higher than 2.6 percent average global growth and the 4 percent growth of developing economies.

Stepping into the new era, China's principal contradiction is between unbalanced and inadequate development and the people's ever-growing needs for a better life. No mention of any target to double GDP will ensure the country can focus on implementing its new development concepts of innovation, coordination, greening, opening up and inclusiveness.

Although economic growth at certain level will still be necessary, freeing itself from any speed constraint will make it easier for China to balance its development goals.

Socialist modernization will be "basically realized" from 2020 to 2035. From 2035 to the middle of the century, China will become a great modern socialist country that is prosperous, strong, democratic, culturally advanced, harmonious and beautiful, according to the report.

With more focus on quality and efficiency than on speed, China will march toward those goals and bring about happiness to the people, and rejuvenation for the Chinese nation.

10.2 Total Quality Management

10.2.1 Quality Laws and Regulations

10.2.1.1 ISO Standards

The International Organization for Standardization (ISO) created the Quality Management System (QMS)[9] standards in 1987. They were the ISO 9000: 1987 series of standards comprising ISO 9001: 1987, ISO 9002: 1987 and ISO 9003: 1987; which were applicable in different types of industries, based on the type of activity or process: designing, production or service delivery.

The standards are reviewed every few years by the International Organization for Standardization. The version in 1994 was called the ISO 9000: 1994 series; consisting of the ISO 9001: 1994, 9002: 1994 and 9003: 1994 versions.

The last major revision was in the year 2000 and the series was called ISO 9000: 2000 series. The ISO 9002 and 9003 standards were integrated into one single certifiable standard: ISO 9001: 2000. After December 2003, organizations holding ISO 9002 or 9003 standards had to complete a transition to the new standard.

ISO released a minor revision, ISO 9001: 2008 on 14 October 2008. It contains no new requirements. Many of the changes were to improve consistency in grammar, facilitating translation of the standard into other languages for use by over 950,000 certified organization in the 175 countries (as at Dec 2007) that use the standard.

The ISO 9004: 2009 document gives guidelines for performance improvement over and above the basic standard (ISO 9001: 2000). This standard provides a measurement framework for improved quality management, similar to and based upon the measurement framework for process assessment.

The Quality Management System standards created by ISO are meant to certify the processes and the system of an organization, not the product or service itself. ISO 9000 standards do not certify the quality of the product or service.

In 2005 the International Organization for Standardization released a standard, ISO 22000, meant for the food industry. This standard covers the values and principles of ISO 9000 and the HACCP standards. It gives one single integrated standard for the food industry and is expected to become more popular in the coming years in such industry.

ISO has also released standards for other industries. For example, Technical Standard TS 16949 defines requirements in addition to those in ISO 9001: 2008 specifically for the automotive industry.

ISO has a number of standards that support quality management. One group describes processes (including ISO/IEC 12207 and ISO/IEC 15288) and another describes process assessment and improvement ISO 15504.

CMMI and IDEAL methods [edit]

The Software Engineering Institute has its own process assessment and improvement methods, called CMMI (Capability Maturity Model Integration) and IDEAL respectively.

Capability Maturity Model Integration (CMMI) is a process improvement training and appraisal program and service administered and marketed by Carnegie Mellon University and required by many DOD and U. S. Government contracts, especially in software development. Carnegie Mellon University claims CMMI can be used to guide process improvement across a project, division, or an entire organization. Under the CMMI methodology, processes are rated according to their maturity levels, which are defined as: Initial, Managed, Defined, Quantitatively Managed, Optimizing. Currently supported is CMMI Version 1. 3. CMMI is registered in the U. S. Patent and Trademark Office by Carnegie Mellon University.

Three constellations of CMMI are:

(1) Product and service development (CMMI for Development);

(2) Service establishment, management, and delivery (CMMI for Services);

(3) Product and service acquisition (CMMI for Acquisition).

CMMI Version 1. 3 was released on November 1, 2010. This release is noteworthy because it updates all three CMMI models (CMMI for development, CMMI for Services, and CMMI for acquisition) to make them consistent and to improve their high maturity practices. The CMMI product team has reviewed more than 1, 150 change requests for the models and 850 for the appraisal method.

As part of its mission to transition mature technology to the software community, the SEI has transferred CMMI - related products and activities to the CMMI institute, a 100% - controlled subsidiary of carnegie innovations, Carnegie Mellon University's technology commercialization enterprise.

10. 2. 1. 2 Establish an Effective Quality System

The quality system refers to the organizational structure, procedures, processes and resources required to implement quality management. In order to achieve its specified quality policy and quality objectives, an enterprise needs to decompose its product quality formation process, set up the necessary organizational structure, clarify the responsibility system, equip the necessary equipment and personnel, and adopt appropriate control methods to affect product quality The various factors of technology, management and personnel are controlled to reduce, eliminate, and especially prevent the occurrence of quality defects. The sum of all these items is the quality system, or the quality system is an organic complex of all these items. It can be seen from the figure below that the establishment of a quality system is the core task of total quality management. Without the quality system, total quality management becomes an empty shell. From this point of view, it is necessary for enterprises to establish a comprehensive quality system, which is the fundamental guarantee for the realization of comprehensive quality management. Figure 10-3 shows TQM factor chart.

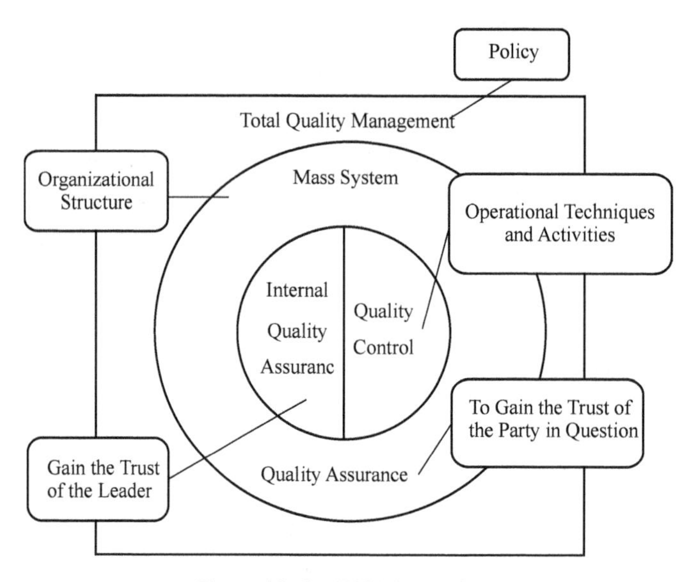

Figure 10-3 TQM factor chart

10. 2. 2 Total Quality Management

At the end of the 1950s, feigenbaum of general electric company of the United States and quality management expert Juran proposed the concept of "Total Quality Management" (TQM), and believed that "Total Quality Management is for the most economical at a high level, and taking into account the conditions of fully satisfying customer requirements for production and providing services, it constitutes an effective system that integrates various departments of the enterprise in the activities of quality development, quality maintenance and quality improvement." The early 1960s, some US companies based on management behavior scientific theory, enterprise quality management to carry out relying on workers "in the self-control" and "defect-free movement" (zero defects), Japan's industrial enterprises to carry out the quality management group (QCCircle/quality control circle) activity line, so that total quality management activities develop rapidly.

The basic method of total quality management can be summarized in four sentences and eighteen words, that is, one process, four stages, eight steps, mathematical statistics methods.

A process, that is, business management is a process. Enterprises should complete different tasks at different times. Every production and operation activity of an enterprise has a process of generation, formation, implementation and verification.

Four stages, according to the theory of management is a process, the United States Daiming Boshi apply it to quality management in the past, concludes that "plan (plan) - execute (do) - check (check) - processing (act) " four The cycle of phases, referred to as PDCA cycle, also known as "Deming cycle" .

Eight steps, in order to solve and improve quality problems, the four phases in the PDCA cycle can also be specifically divided into eight steps. 1) Planning stage: Analyze the status quo to find out the existing quality problems; analyze the various causes or influencing factors of the quality problems; find out the main factors that affect the quality; propose plans and formulate measures for the main factors that affect the quality. 2) Implementation stage: implement the plan and implement the measures. 3) Inspection stage: inspect the implementation of the plan. 4) Processing stage: Summarize experience, consolidate achievements, and standardize work results; raise unresolved problems and move to the next cycle.

When applying PDCA's four cycle stages and eight steps to solve quality problems, it is necessary to collect and sort out a large number of books and materials and use scientific methods for systematic analysis. The seven most commonly used statistical methods are Pareto Diagrams, Cause and Effect Diagrams, Histograms, Stratification, Correlation Diagrams, Control Charts and Statistical Analysis Tables. This method is based on mathematical statistics, which is not only scientific and reliable, but also relatively intuitive.

10. 2. 3 "PDCA Cycle"

The current effective is the PDCA cycle work process commonly used by enterprises to implement total quality management. The basic content of the "PDCA cycle" process is to make a plan before doing something and then implement it according to the plan, and check and adjust during the implementation process, and summarize and deal with it when the plan is completed. The American Deming summed up this rule as the "PDCA cycle", as shown in the figure, PDCA stands for the four words of Plan, Do, Check, and Action in English. It reflects the need for quality management. The four stages to follow.

Phase P: Find out to meet the requirements of users, and to achieve the most economical effect as the goal, through investigation, design, trial production, formulation of technical and economic indicators, quality goals, management projects and specific measures and methods to achieve these goals. This is the planning stage.

Stage D: It is to implement in accordance with the plans and measures formulated. This is the execution phase.

Phase C: It is to check the implementation and effects against the plan and discover the experience and problems in the plan implementation process in time. This is the inspection phase.

Stage A: It is to take measures based on the results of the inspection, consolidate achievements, learn lessons, and fight again. This is the summary processing stage.

These four stages can be roughly divided into eight steps (see the Figure 10-4 below).

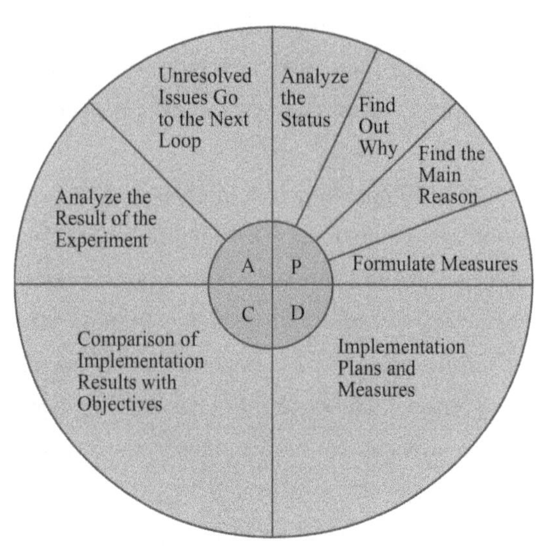

扫一扫查看彩图

Figure 10-4 PDCA cycle

词汇

生词	音标	释义
implementing	[ˈɪmplɪmentɪŋ]	*v.* 使生效；贯彻；执行；实施
supervise	[ˈsuːpəvaɪz]	*v.* 监督；管理；指导；主管
executive	[ɪgˈzekjətɪv]	*n.* （公司或机构的）经理，主管领导，管理人员；（统称公司或机构的）行政领导，领导层；（政府的）行政部门
unattainable	[ˌʌnəˈteɪnəbl]	*adj.* 无法得到的；难以达到的
daunting	[ˈdɔːntɪŋ]	*adj.* 使人畏惧的；令人胆怯的；让人气馁的 *v.* 使胆怯；使气馁；使失去信心
emerged	[ɪˈmɜːdʒd]	*v.* （从隐蔽处或暗处）出现，浮现，露出；暴露；露出真相；被知晓；露头；显现；显露
overcapacity	[ˌəʊvəkəˈpæsətɪ]	*n.* 生产能力过剩
sophisticated	[səˈfɪstɪkeɪtɪd]	*adj.* 见多识广的；老练的；见过世面的；复杂巧妙的；先进的；精密的；水平高的；在行的
corruption	[kəˈrʌpʃn]	*n.* 腐败；贪污；贿赂；受贿；使人堕落的行为；腐蚀；（单词或短语的）变体
deductive	[dɪˈdʌktɪv]	*adj.* 演绎的；推论的；推理的
inductive	[ɪnˈdʌktɪv]	*adj.* 归纳法的；归纳的；电感应的

长难句

"Scientific management" puts forward the requirement of scientific division of labor among personnel, and separates the planning function from the executive function, and adds an inspection link in the middle to supervise and inspect the implementation of the plan, design, product standards and other items.

《科学管理》提出了劳动力科学分工的要求，将计划职能与执行职能分开，在中间增加了检查环节，对计划、设计、产品标准等的执行情况进行监督检查。

Reference

[1] Deming W E. The New Economics for Industry, Government, and Education [M]. Boston: MIT Press, 1993.

[2] Deming W E. Out of the crisis [M]. London: MIT Press, 2000.

[3] Kolesar P J. What Deming Told the Japanese in 1950 [J]. Quality Management Journal, 1994, 2 (1): 9-24.

[4] Rodd L R. The Art of Chairing: What Deming Taught the Japanese and the Japanese Taught Me [J]. Adfl Bulletin, 2001, 32: 5-9.

[5] Phillips-Donaldson Debbie. 100 Years of Juran [J]. Quality Progress, Milwaukee, Wisconsin: American Society for Quality, 2004, 37 (5): 25-39.

[6] Joseph M. Jump up to Juran [M]. Architect of Quality: The Autobiography of Dr. Joseph M. Juran (1 ed.), 2004.

[7] Nick. Jump up to Bunkley [N]. The New York Times, 2008-08-21.

[8] Philip C. Developer of the Zero-Defects Concept [N]. The New York Times, 2012-09-01.

[9] ISO 9001 Quality Management System QMS Certification. Indian Register Quality Systems [S], 2008.

11 Management Information System

11.1 Information System and Management

A Management Information System (MIS) is an information system[1] used for decision-making, and for the coordination, control, analysis, and visualization of information in an organization[1].

Management Information Systems (MIS) is the study of people, technology, organizations, and the relationships among them. MIS professionals help firms realize maximum benefit from investment in personnel, equipment, and business processes. MIS is a people-oriented field with an emphasis on service through technology.

The study of the management information systems involves people, processes and technology in an organizational context[2-3].

In a corporate setting, the ultimate goal of the use of a management information system is to increase the value and profits of the business.

This is done by providing managers with timely and appropriate information allowing them to make effective decisions within a shorter period of time[4].

While it can be contested that the history of management information systems date as far back as companies using ledgers to keep track of accounting, the modern history of MIS can be divided into five eras originally identified by Kenneth C. Laudon and Jane Laudon in their seminal textbook Management Information Systems[5-6].

First Era—Mainframe and minicomputer computing;

Second Era—Personal computers;

Third Era—Client/server networks;

Fourth Era—Enterprise computing;

Fifth Era—Cloud computing.

The first era (mainframe and minicomputer computing) was ruled by IBM and their mainframe computers for which they supplied both the hardware and software. These computers would often take up whole rooms and require teams to run them. As technology advanced, these computers were able to handle greater capacities and therefore reduce their cost. Smaller, more affordable minicomputers allowed larger businesses to run their own computing centers in-house / on-site / on-premises.

The second era (personal computers) began in 1965 as microprocessors started to compete with mainframes and minicomputers and accelerated the process of decentralizing computing power from large data centers to smaller offices. In the late 1970s, minicomputer technology gave way to personal computers and relatively low-cost computers were becoming mass market commodities, allowing businesses to provide their employees access to computing power that ten years before would have cost tens of thousands of dollars. This proliferation of computers created a ready market for interconnecting networks and the popularization of the Internet. (The first microprocessor — a four-bit device intended for a programmable calculator — was introduced in 1971 and microprocessor-based systems were not readily available for several years. The MITS Altair 8800 was the first commonly known microprocessor-based system, followed closely by the Apple Ⅰ and Ⅱ. It is arguable that the microprocessor-based system did not make significant inroads into minicomputer use until 1979, when VisiCalc prompted record sales of the Apple Ⅱ on which it ran. The IBM PC introduced in 1981 was more broadly palatable to business, but its limitations gated its ability to challenge minicomputer systems until perhaps the late 1980s to early 1990s.)

The third era (client/server networks) arose as technological complexity increased, costs decreased, and the end-user (now the ordinary employee) required a system to share information with other employees within an enterprise. Computers on a common network shared information on a server. This lets thousands and even millions of people access data simultaneously on networks referred to as Intranets.

The fourth era (enterprise computing) enabled by high speed networks, consolidated the original department specific software applications into integrated software platforms referred to as enterprise software. This new platform tied all aspects of the business enterprise together offering rich information access encompassing the complete management structure.

11.1.1 Concept of Information

Information can be thought of as the resolution of uncertainty; it is that which answers the question of "What an entity is" and thus defines both its essence and nature of its characteristics. The concept of information has different meanings in different contexts[7].

Thus, the concept becomes related to notions of constraint, communication, control, data, form, education, knowledge, meaning, understanding, mental, stimuli, pattern, perception, representation, and entropy.

Information is associated with data, as data represent values attributed to parameters, and information is data in context and with meaning attached. Information also relates to knowledge, as knowledge signifies understanding of an abstract or concrete concept.

In terms of communication, information is expressed either as the content of a message or through direct or indirect observation. That which is perceived can be construed as a message in its own right, and in that sense, information is always conveyed as the content of a message.

Information can be encoded into various forms for transmission and interpretation (for example, information may be encoded into a sequence of signs, or transmitted via a signal). It can also be encrypted for safe storage and communication.

The uncertainty of an event is measured by its probability of occurrence and is inversely proportional to that. The more uncertain an event, the more information is required to resolve uncertainty of that event. The bit is a typical unit of information, but other units such as the nat may be used. For example, the information encoded in one "fair" coin flip is $\log_2(2/1) = 1$ bit, and in two fair coin flips is $\log_2(4/1) = 2$ bits.

11.1.2 Concept and Development of Information System

An **Information System** (**IS**) is a formal, sociotechnical, organizational system designed to collect, process, store, and distribute information[8].

In a sociotechnical perspective, information systems are composed by four components: task, people, structure (or roles), and technology[9].

Information system, an integrated set of components for collecting, storing, and processing data and for providing information, knowledge, and digital products. Business firms and other organizations rely on information systems to carry out and manage their operations, interact with their customers and suppliers, and compete in the marketplace. Information systems are used to run interorganizational supply chains and electronic markets. For instance, corporations use information systems to process financial accounts, to manage their human resources, and to reach their potential customers with online promotions. Many major companies are built entirely around information systems. These include eBay, a largely auction marketplace; Amazon, an expanding electronic mall and provider of cloud computing services; Alibaba, a business-to-business e-marketplace; and Google, a search engine company that derives most of its revenue from keyword advertising on Internet searches. Governments deploy information systems to provide services cost-effectively to citizens. Digital goods—such as electronic books, video products, and software—and online services, such as gaming and social networking, are delivered with information systems. Individuals rely on information systems, generally Internet-based, for conducting much of their personal lives: for socializing, study, shopping, banking, and entertainment.

A **computer information system** is a system composed of people and computers that processes or interprets information. The term is also sometimes used to simply refer to a computer system with software installed[10-13].

While a computer is an inherently diverse tool, businesses generally use a group of networked computers to collect, organize, store, and transmit information. This network is also known as a computer information system. In the field of computer information systems, professionals work to optimize the application of networked computers in business environments.

To be effective in this effort, these professionals must learn how to improve business processes by implementing a computer information system that can accommodate the specific needs of their organization. For example, if an organization is concerned with the productivity of its employees, IT professionals could use the existing computer information system to track and measure relevant metrics. The data from such a system could then be used to design workplace policies that better promote optimal use of labor hours.

Information systems is an academic study of systems with a specific reference to information and the complementary networks of hardware and software that people and organizations use to collect, filter, process, create and also distribute data. An emphasis is placed on an information system having a definitive boundary, users, processors, storage, inputs, outputs, and the aforementioned communication networks[14].

Any specific information system aims to support operations, management, and decision-making. An information system is the information and communication technology (ICT) that an organization uses, and also the way in which people interact with this technology in support of business processes[15-17].

Some authors make a clear distinction between information systems, computer systems, and business processes. Information systems typically include an ICT component but are not purely concerned with ICT, focusing instead on the end-use of information technology. Information systems are also different from business processes. Information systems help to control the performance of business processes[18].

Alter argues for advantages of viewing an information system as a special type of work system. A work system is a system in which humans or machines perform processes and activities using resources to produce specific products or services for customers. An information system is a work system whose activities are devoted to capturing, transmitting, storing, retrieving, manipulating, and displaying information[19-21].

As such, information systems inter-relate with data systems on the one hand and activity systems on the other. An information system is a form of communication system in which data represent and are processed as a form of social memory. An information system can also be considered a semi-formal language which supports human decision making and action.

Information systems are the primary focus of study for organizational informatics[22].

11.2 Introduction to Management Information System

Information management (IM) concerns a cycle of organizational activity: the acquisition of information from one or more sources, the custodianship and the distribution of that information to those who need it, and its ultimate disposition through archiving or deletion.

This cycle of information organization involves a variety of stakeholders, including those who are responsible for assuring the quality, accessibility and utility of acquired information; those who are responsible for its safe storage and disposal; and those who need it for decision making. Stakeholders might have rights to originate, change, distribute or delete information according to organizational information management policies.

Information management embraces all the generic concepts of management, including the planning, organizing, structuring, processing, controlling, evaluation and reporting of information activities, all of which is needed in order to meet the needs of those with organizational roles or functions that depend on information. These generic concepts allow the information to be presented to the audience or the correct group of people. After individuals are able to put that information to use, it then gains more value.

Information management is closely related to, and overlaps with, the management of data, systems, technology, processes and – where the availability of information is critical to organizational success – *strategy*. This broad view of the realm of information management contrasts with the earlier, more traditional view, that the life cycle of managing information is an operational matter that requires specific procedures, organizational capabilities and standards that deal with information as a product or a service.

In the 1970s, the management of information largely concerned matters closer to what would now be called data management: punched cards, magnetic tapes and other record-keeping media, involving a life cycle of such formats requiring origination, distribution, backup, maintenance and disposal. At this time the huge potential of information technology began to be recognized: for example a single chip storing a whole book, or electronic mail moving messages instantly around the world, remarkable ideas at the time[23].

With the proliferation of information technology and the extending reach of information systems in the 1980s and 1990s, information management took on a new form. Progressive businesses such as British Petroleum transformed the vocabulary of what was then "IT management", so that "systems analysts" became "business analysts", "monopoly supply" became a mixture of "insourcing" and "outsourcing", and the large IT function was transformed into "lean teams" that began to allow some agility in the processes that harness information for business benefit. The scope of senior management interest in information at British Petroleum extended from the creation of value through improved business processes, based upon the effective management of information, permitting the implementation of appropriate information systems (or "applications") that were operated on IT infrastructure that was outsourced. In this way, information management was no longer a simple job that could be performed by anyone who had nothing else to do, it became highly strategic and a matter for senior management attention. An understanding of the technologies involved, an ability to manage information systems projects and business change well, and a willingness to align technology and business strategies all became necessary.

11.2.1 Concept of Management Information System

The MIS is an idea which is associated with man, machine, marketing, and methods for collecting information's from the internal and external source and processing this information for the purpose of facilitating the process of decision-making of the business.

MIS is not new, only the computerization is new, before computers MIS techniques existed to supply managers with the information that would permit them to plan and control business operations. The computer has added on more dimensions such as speed, accuracy and increased volume of data that permit the consideration of more alternatives in decision-making process.

Management information system is an integrated set of component or entities that interact to achieve a particular function, objective, or goal. Therefore, it is a computer-based system that provides information for decisions making on planning, organizing, and controlling the operation of the sub-system of the firm and provides a synergistic organization in the process.

The component of an information system includes: a hardware which is used for input/output process and storage of data, software used to process data and also to instruct the hand-ware component, data bases which is the location in the system where all the organization data will be automated and procedures which is a set of documents that explain the structure of that management information system.

There are various driving factors of management information system for example:

Technological revolutions in all sectors make modern managers to need to have access to large amount of selective information for the complex tasks and decisions.

The lifespan of most product has continued getting shorter and shorter and therefore the challenge to the manager is to design product that will take a longer shelf life and in order to do this, the manager must be able to keep abreast of the factors that influences the organization product and services thus, management information system come in handy in supporting the process.

There is huge amount of information available to today's manager and this had therefore meant that managers are increasingly relying on management information system to access the exploding information. Management information services helps manager to access relevant, accurate, up-to-date information which is the surer way of making accurate decisions. It also helps in automation and incorporation of research and management science techniques into the overall management information system for example probability theory.

The management information services are capable of taking advantage of the computational ability of the company like processing, storage capacity among others.

Based on this relevancy, management information system should be installed and upgraded in various organizations since today's managers need them to access information for **managerial** decision making and also management functions.

11.2.2 Classification of Management Information System

11.2.2.1 Classification of Management Information System

Transaction processing is a way of computing that divides work into individual, indivisible operations, called transactions. A transaction processing system (TPS) is a software system, or software/hardware combination, that supports transaction processing.

The first transaction processing system was SABRE, made by IBM for American Airlines, which became operational in 1970. Designed to process up to 83,000 transactions a day, the system ran on two IBM 7090 computers. SABRE was migrated to IBM system/360 computers in 1972, and became an IBM product first as Airline Control Program (ACP) and later as Transaction Processing Facility (TPF). In addition to airlines TPF is used by large banks, credit card companies, and hotel chains.

The hewlett-packard non stop system (formerly tandem non stop) was a hardware and software system designed for Online Transaction Processing (OLTP) introduced in 1976. The systems were designed for transaction processing and provided an extreme level of availability and data integrity.

TPS processes transaction and produces reports. It represents the automation of the fundamental, routine processing used to support business operations. It does not provide any information to the user to his/her decision-making. TPS uses data and produces data as shown in the following diagram.

Previously, TPS was known as Management Information System. Prior to computers, data processing was performed manually or with simple machines. The domain of TPS is at the lowest level of the management **hierarchy** of an organization.

11.2.2.2 Management Information System (MIS)

MIS is an information system, which processes data and converts it into information. A management information system uses TPS for its data inputs. The information generated by the information system may be used for control of operations, strategic and long-range planning. Short-range planning, management control, and other managerial problem solving. It encompasses processing in support of a wide range of organizational functions & management processes. MIS is capable of providing analysis, planning & decision-making support. The functional areas of a business may be marketing, production, human resource, finance, and accounting.

Businesses use information systems at all levels of operation to collect, process, and store data. Management aggregates and disseminates this data in the form of information needed to carry out the daily operations of business. Everyone who works in business, from someone who pays the bills to the person who makes employment decisions, uses information systems. A car dealership could use a computer database to keep track of which products sell best. A retail store might use a computer-based information system to sell products over the Internet. In fact, many (if not most) businesses concentrate on the alignment of MIS with business goals to achieve competitive advantage over other businesses.

MIS professionals create information systems for data management (i.e., storing, searching, and analyzing data). In addition, they manage various information systems to meet the needs of managers, staff, and customers. By working collaboratively with various members of their work group, as well as with their customers and clients, MIS professionals are able to play a key role in areas such as information security, integration, and exchange. As an MIS major, you will learn to design, implement, and use business information systems in innovative ways to increase the effectiveness and efficiency of your company.

11.2.2.3 Decision Support System (DSS)

A decision support system (DSS) is an information system application that assists decision-making. DSS tends to be used in planning, analyzing alternatives, and trial and error search for solution. The elements of the decision support system include a database, model base & software. The main application areas of DSS are production, finance, and marketing.

Typical information used by a DSS includes target or projected revenue, sales figures or past ones from different time periods, and other inventory- or operations-related data.

A decision support system gathers and analyzes data, synthesizing it to produce comprehensive information reports. In this way, as an informational application, a DSS differs from an ordinary operations application, whose function is just to collect data.

The DSS can either be completely computerized or powered by humans. In some cases, it may combine both. The ideal systems analyze information and actually make decisions for the user. At the very least, they allow human users to make more informed decisions at a quicker pace. Figure 11-1 shows elements of DSS.

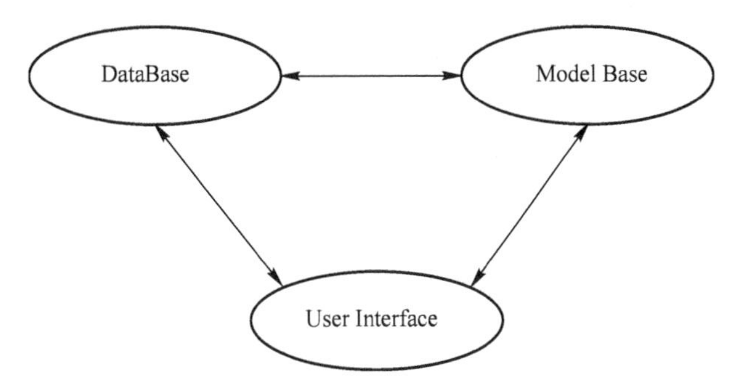

Figure 11-1 Elements of DSS

DSS can be differentiated from MIS on the basis of processing the information. MIS processes data to convert it into information. DSS processes information to support the decision-making process of a manager.

11.2.2.4 Executive Support System (ESS)

Executive Support System (ESS) is an extension of the management information system, which is a special kind of DSS; An ESS is specially tailored for the use of chief executive of an organization to support his decision-making. It includes various types of decision-making, but it is more specific and person oriented.

EIS emphasizes graphical displays and easy-to-use user interfaces. They offer strong reporting and drill-down capabilities. In general, EIS are enterprise-wide DSS that help top-level executives analyze, compare, and highlight trends in important variables so that they can monitor performance and identify opportunities and problems. EIS and data warehousing technologies are converging in the marketplace.

Traditionally, executive information systems were mainframe computer-based programs. The purpose was to package a company's data and to provide sales performance or market research statistics for decision makers, such as, marketing directors, chief executive officer, who were not necessarily well acquainted with computers. The objective was to develop computer applications that highlighted information to satisfy senior executives' needs. Typically, an EIS provides only data that supported executive level decisions, not all company data.

Today, the application of EIS is not only in typical corporate hierarchies, but also at lower corporate levels. As some client service companies adopt the latest enterprise information systems, employees can use their personal computers to get access to the company's data and identify information relevant to their decision making. This arrangement provides relevant information to upper and lower corporate levels.

11.2.2.5 Office Automation Systems (OAS)

Office automation refers to the application of computes and communication technology to office functions. Office automation systems are meant to improve the productivity of managers at various levels of management of providing secretarial assistance and better communication facilities.

The backbone of office automation is a LAN, which allows users to transfer data, mail and even voice across the network. All office functions, including dictation, typing, filing, copying, fax, Telex, microfilm and records management, telephone and telephone switchboard operations, fall into this category. Office automation was a popular term in the 1970s and 1980s as the desktop computer exploded onto the scene.

Advantages are:

(1) Office automation can get many tasks accomplished faster.

(2) It eliminates the need for a large staff.

(3) Less storage is required to store data.

(4) Multiple people can update data simultaneously in the event of changes in schedule.

Office activities may be grouped under two classes, namely I) Activities performed by clerical personnel (clerks, secretaries, typist, etc.,) and Ⅱ) Activities performed by the executives (managers, engineers or other professionals like economist, research etc.

In the first category, the following is a list of activities.

(1) Typing;

(2) Mailing;

(3) Scheduling of meetings and conferences;

(4) Calendar keeping;

(5) Retrieving documents.

The following is a list of activities in the second category (managerial category):

(1) Conferencing;

(2) Production of information (messages, memos, reports, etc.) and controlling performance.

Business expert systems: These systems are one of the main types of knowledge - based information systems. These systems are based on artificial intelligence and are advanced information systems. A business expert system is a knowledge based on information system that uses its knowledge about a specific, complex application area to act as an expert. The main components of an expert system are:

(1) Knowledge base;

(2) Interface engine;

(3) User interface.

11.2.3 Manufacturing Resource Planning (MRP Ⅱ)

Manufacturing resource planning (MRP Ⅱ) is defined as a method for the effective planning of all resources of a manufacturing company. Ideally, it addresses operational planning in units, financial planning, and has a simulation capability to answer "what-if" questions and extension of closed-loop MRP.

This is not exclusively a software function, but the management of people skills, requiring a dedication to database accuracy, and sufficient computer resources. It is a total company management concept for using human and company resources more productively.

MRP Ⅱ is not a proprietary software system and can thus take many forms. It is almost impossible to visualize an MRP Ⅱ system that does not use a computer, but an MRP Ⅱ system can be based on either purchased-licensed or in-house software.

Almost every MRP Ⅱ system is modular in construction. Characteristic basic modules in an MRP Ⅱ system are:

(1) Master production schedule (MPS);

(2) Item master data (technical data);

(3) Bill of materials (BOM) (technical data);

(4) Production resources data (manufacturing technical data);

(5) Inventories and orders (inventory control);

(6) Purchasing management;

(7) Material requirements planning (MRP);

(8) Shop floor control (SFC);

(9) Capacity planning or capacity requirements planning (CRP);

(10) Standard costing (cost control) and frequently also actual or FIFO costing, and weighted average costing;

(11) Cost reporting / management (cost control).

Together with auxiliary systems such as:

(1) Business planning;

(2) Lot traceability;

(3) Contract management;

(4) Tool management;

(5) Engineering change control;

(6) Configuration management;

(7) Shop floor data collection;

(8) Sales analysis and forecasting;

(9) Finite capacity scheduling (FCS).

And related systems such as:

(1) General ledger;

(2) Accounts payable (purchase ledger);

(3) Accounts receivable (sales ledger);

(4) Sales order management;

(5) Distribution resource planning (DRP);

(6) Automated warehouse management;

(7) Project management;

(8) Technical records;

(9) Estimating;

(10) Computer-aided design/computer-aided manufacturing (CAD/CAM);

(11) CAPP.

The MRP Ⅱ system integrates these modules together so that they use common data and freely exchange information, in a model of how a manufacturing enterprise should and can operate. The MRP Ⅱ approach is therefore very different from the "point solution" approach, where individual systems are deployed to help a company plan, control, or manage a specific activity. MRP Ⅱ is by definition fully integrated or at least fully interfaced.

MRP Ⅱ systems can provide:

(1) Better control of inventories;

(2) Improved scheduling;

(3) Productive relationships with suppliers.

For design / engineering:

(1) Improved design control;

(2) Better quality and quality control.

For financial and costing:

(1) Reduced working capital for inventory;

(2) Improved cash flow through quicker deliveries;

(3) Accurate inventory records.

Authors like Pochet and Wolsey argue that MRP and MRP Ⅱ, as well as the planning modules in current APS and ERP systems, are actually sets of heuristics. Better production plans could be obtained by optimization over more powerful mathematical programming models, usually integer programming models. While they acknowledge that the use of heuristics, like those prescribed by MRP and MRP Ⅱ, were necessary in the past due to lack of computational power to solve complex optimization models, this is mitigated to some extent by recent improvements in computers[24].

11.2.3.1 MRP and MRP II

Material requirements planning (MRP) and manufacturing resource planning (MRP Ⅱ) are predecessors of enterprise resource planning (ERP), a business information integration system. The development of these manufacturing coordination and integration methods and tools made today's ERP systems possible. Both MRP and MRP Ⅱ are still widely used, independently and as modules of more comprehensive ERP systems, but the original vision of integrated information systems as we know them today began with the development of MRP and MRP Ⅱ in manufacturing.

MRP (and MRP Ⅱ) evolved from the earliest commercial database management package developed by Gene Thomas at IBM in the 1960s. The original structure was called BOMP (bill-of-materials processor), which evolved in the next generation into a more generalized tool called DBOMP (Database Organization and Maintenance Program). These were run on mainframes, such as IBM/360.

The vision for MRP and MRP Ⅱ was to centralize and integrate business information in a way that would facilitate decision making for production line managers and increase the efficiency of the production line overall. In the 1980s, manufacturers developed systems for calculating the resource requirements of a production run based on sales forecasts. In order to calculate the raw materials needed to produce products and to schedule the purchase of those materials along with the machine and labor time needed, production managers recognized that they would need to use computer and software technology to manage the information. Originally, manufacturing operations built custom software programs that ran on mainframes.

Material requirements planning (MRP) was an early iteration of the integrated information systems vision. MRP information systems helped managers determine the quantity and timing of raw materials purchases. Information systems that would assist managers with other parts of the manufacturing process, MRP Ⅱ, followed. While MRP was primarily concerned with materials, MRP Ⅱ was concerned with the integration of all aspects of the manufacturing process, including materials, finance, and human resources.

Like today's ERP systems, MRP Ⅱ was designed to tell us about a lot of information by way of a centralized database. However, the hardware, software, and relational database technology of the 1980s was not advanced enough to provide the speed and capacity to run these systems in real-time, and the cost of these systems was prohibitive for most businesses. Nonetheless, the vision had been established, and shifts in the underlying business processes along with rapid advances in technology led to the more affordable enterprise and application integration systems that big businesses and many medium and smaller businesses use today.

11.2.3.2 General Concepts

Material requirements planning (MRP), and manufacturing resource planning (MRP Ⅱ) are both incremental information integration business process strategies that are implemented using hardware and modular software applications linked to a central database that stores and delivers business data and information.

MRP is concerned primarily with manufacturing materials while MRP Ⅱ is concerned with the coordination of the entire manufacturing production, including materials, finance, and human resources. The goal of MRP Ⅱ is to provide consistent data to all members in the manufacturing process as the product moves through the production line.

Paper-based information systems and non-integrated computer systems that provide paper or disk outputs result in many information errors, including missing data, redundant data, numerical errors that result from being incorrectly keyed into the system, incorrect calculations based on numerical errors, and bad decisions based on incorrect or old data. In addition, some data is unreliable in non-integrated systems because the same data is categorized differently in the individual databases used by different functional areas. Figure 11-2 shows the process of MRP Ⅱ.

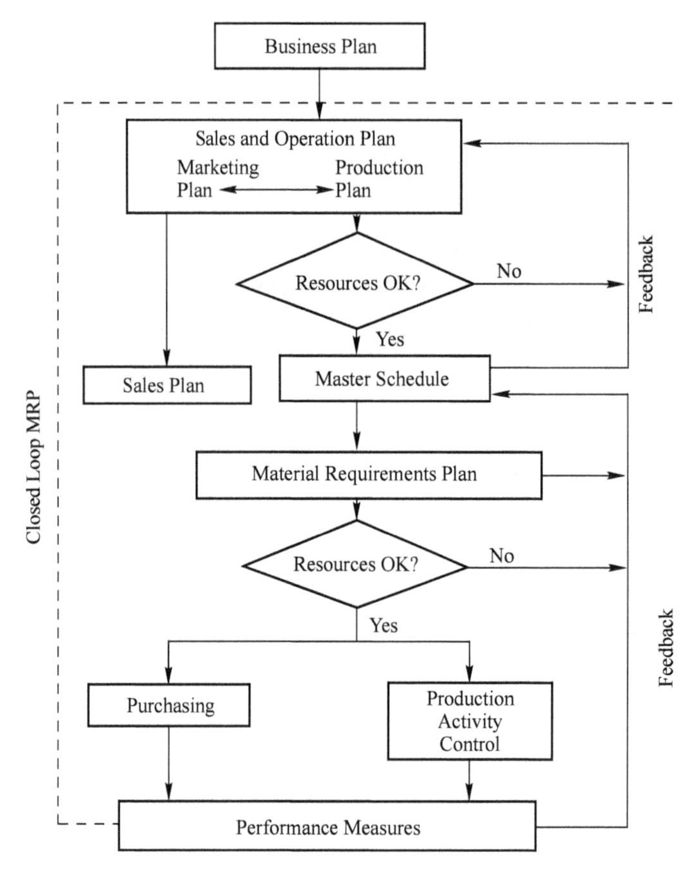

Figure 11-2 MRP Ⅱ

MRP Ⅱ systems begin with MRP, material requirements planning. MRP allows for the input of sales forecasts from sales and marketing, or of actual sales demand in the form of customers' orders. These demands determine the raw materials demand. MRP and MRP Ⅱ systems draw on a master production schedule, the breakdown of specific plans for each product on a line. **While MRP allows for the coordination of raw materials purchasing, MRP Ⅱ facilitates the development of a detailed production schedule that accounts for machine and labor capacity, scheduling the production runs according to the arrival of materials.** An MRP Ⅱ output is a final labor and machine schedule. Data about the cost of production, including machine time, labor time and materials used, as well as final production numbers, is provided from the MRP Ⅱ system to accnting and finance.

For the companies that want to integrate their other departments with their manufacturing management, ERP software are necessary.

11.2.3.3 Enterprise Resource Planning (ERP)

Enterprise resource planning (ERP) is the integrated management of main business processes, often in real time and mediated by software and technology.

ERP is usually referred to as a category of business management software—typically a suite of integrated applications—that an organization can use to collect, store, manage, and interpret data from many business activities.

ERP provides an integrated and continuously updated view of core business processes using common databases maintained by a database management system. ERP systems track business resources—cash, raw materials, production capacity—and the status of business commitments: orders, purchase orders, and payroll. The applications that make up the system share data across various departments (manufacturing, purchasing, sales, accounting, etc.) that provide the data. ERP facilitates information flow between all business functions and manages connections to outside stakeholders.

Enterprise system software is a multibillion-dollar industry that produces components supporting a variety of business functions. IT investments have, as of 2011, become one of the largest categories of capital expenditure in United States-based businesses. Though early ERP systems focused on large enterprises, smaller enterprises increasingly use ERP systems.

The ERP system integrates varied organizational systems and facilitates error-free transactions and production, thereby enhancing the organization's efficiency. However, developing an ERP system differs from traditional system development. ERP systems run on a variety of computer hardware and network configurations, typically using a database as an information repository.

11.2.4 Enterprise Resource Planning (ERP)

11.2.4.1 Enterprise Resource Planning

Figure 11-3 shows some typical ERP modules.

Enterprise Resource Planning (ERP) is the integrated management of main business processes, often in real time and mediated by software and technology.

ERP is usually referred to as a category of business management software—typically a suite of integrated applications—that an organization can use to collect, store, manage, and interpret data from many business activities.

ERP provides an integrated and continuously updated view of core business processes using common databases maintained by a database management system. ERP systems track business resources—cash, raw materials, production capacity—and the status of business commitments: orders, purchase orders, and payroll. The applications that make up the system share data across various departments (manufacturing, purchasing, sales, accounting, etc.) that provide the data. ERP facilitates information flow between all business functions and manages connections to outside stakeholders.

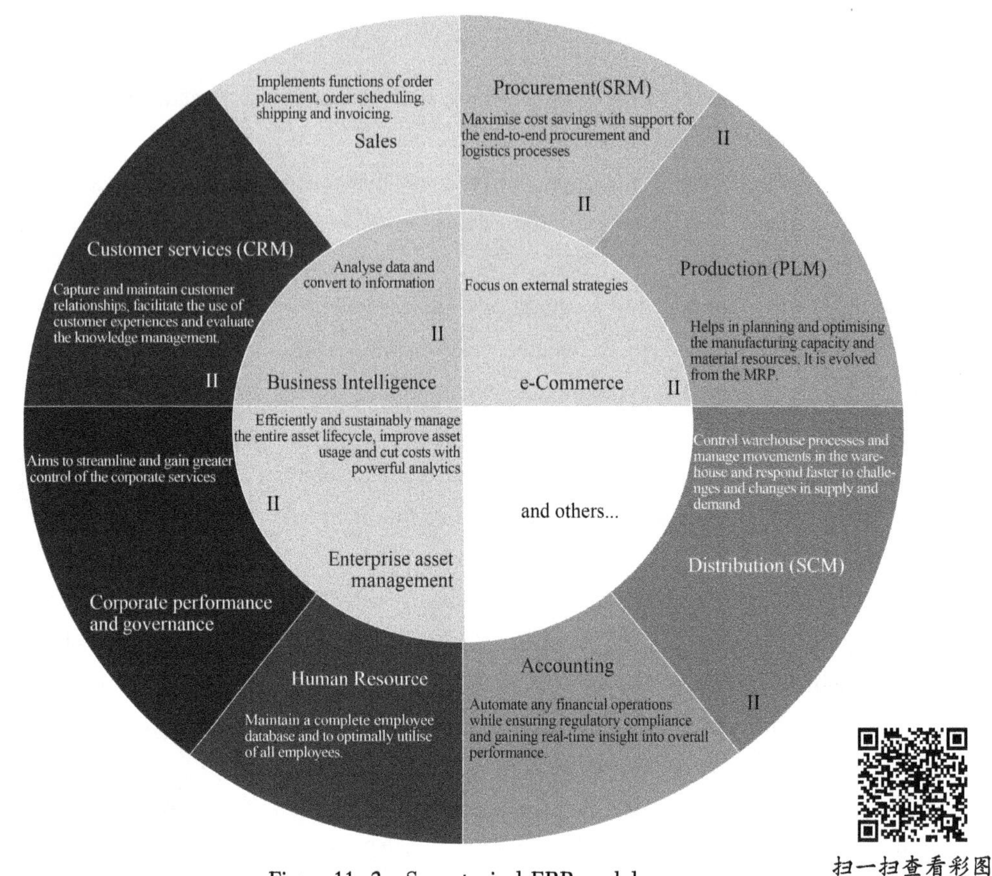

扫一扫查看彩图

Figure 11-3 Some typical ERP modules

Enterprise system software is a multibillion-dollar industry that produces components supporting a variety of business functions. IT investments have, as of 2011, become one of the largest categories of capital expenditure in United States-based businesses. Though early ERP systems focused on large enterprises, smaller enterprises increasingly use ERP systems.

The ERP system integrates varied organizational systems and facilitates error-free transactions and production, thereby enhancing the organization's efficiency. However, developing an ERP system differs from traditional system development. ERP systems run on a variety of computer hardware and network configurations, typically using a database as an information repository.

11.2.4.2 Origin

The gartner group first used the acronym ERP in the 1990s to include the capabilities of Material Requirements Planning (MRP), and the later manufacturing resource planning (MRP Ⅱ), as well as computer-integrated manufacturing. Without replacing these terms, ERP came to represent a larger whole that reflected the evolution of application integration beyond manufacturing.

Not all ERP packages are developed from a manufacturing core; ERP vendors variously began assembling their packages with finance - and - accounting, maintenance, and human - resource components. By the mid - 1990s ERP systems addressed all core enterprise functions. Governments and non-profit organizations also began to use ERP systems.

11. 2. 4. 3 Expansion

ERP systems experienced rapid growth in the 1990s. Because of the year 2000 problem many companies took the opportunity to replace their old systems with ERP.

ERP systems initially focused on automating back office functions that did not directly affect customers and the public. Front office functions, such as customer relationship management (CRM), dealt directly with customers, or e - business systems such as e - commerce, e - government, e-telecom, and e-finance—or supplier relationship management (SRM) became integrated later, when the internet simplified communicating with external parties.

"ERP Ⅱ" was coined in 2000 in an article by gartner publications entitled ERP is dead—long live ERP Ⅱ. It describes web-based software that provides real-time access to ERP systems to employees and partners (such as suppliers and customers). The ERP Ⅱ role expands traditional ERP resource optimization and transaction processing. Rather than just manage buying, selling, etc. ERP Ⅱ leverages information in the resources under its management to help the enterprise collaborate with other enterprises. ERP Ⅱ is more flexible than the first-generation ERP. Rather than confine ERP system capabilities within the organization, it goes beyond the corporate walls to interact with other systems. Enterprise application suite is an alternate name for such systems. ERP Ⅱ systems are typically used to enable collaborative initiatives such as supply chain management (SCM), customer relationship management (CRM), and business intelligence (BI) among business partner organizations through the use of various e-business technologies.

Developers now make more effort to integrate mobile devices with the ERP system. ERP vendors are extending ERP to these devices, along with other business applications. Technical stakes of modern ERP concern integration—hardware, applications, networking, supply chains. ERP now covers more functions and roles—including decision making, stakeholders' relationships, standardization, transparency, globalization, etc.

11. 2. 4. 4 Characteristics

ERP systems typically include the following characteristics:

(1) An integrated system;

(2) Operates in (or near) real time;

(3) A common database that supports all the applications;

(4) A consistent look and feel across modules;

(5) Installation of the system with elaborate application/data integration by the Information Technology (IT) department, provided the implementation is not done in small steps;

(6) Deployment options include: on-premises, cloud hosted, or SaaS.

11. 2. 4. 5 Functional Areas

An ERP system covers the following common functional areas. In many ERP systems, these are called and grouped together as ERP modules:

(1) Financial accounting: general ledger, fixed assets, payables including vouchering, matching and payment, receivables and collections, cash management, financial consolidation.

(2) Management accounting: budgeting, costing, cost management, activity based costing.

(3) Human resources: recruiting, training, rostering, payroll, benefits, retirement and pension plans, diversity management, retirement, separation.

(4) Manufacturing: engineering, bill of materials, work orders, scheduling, capacity, workflow management, quality control, manufacturing process, manufacturing projects, manufacturing flow, product life cycle management.

(5) Order processing: order to cash, order entry, credit checking, pricing, available to promise, inventory, shipping, sales analysis and reporting, sales commissioning.

(6) Supply chain management: supply chain planning, supplier scheduling, product configurator, order to cash, purchasing, inventory, claim processing, warehousing (receiving, put away, picking and packing).

(7) Project management: project planning, resource planning, project costing, work breakdown structure, billing, time and expense, performance units, activity management.

(8) Customer relationship management (CRM): sales and marketing, commissions, service, customer contact, call center support - CRM systems are not always considered part of ERP systems but rather business support systems (BSS).

(9) Data services: various "self - service" interfaces for customers, suppliers and/or employees.

11.2.4.6 Postmodern ERP

The term "postmodern ERP" was coined by Gartner in 2013, when it first appeared in the paper series "Predicts 2014". According to Gartner's definition of the postmodern ERP strategy, legacy, monolithic and highly customized ERP suites, in which all parts are heavily reliant on each other, should sooner or later be replaced by a mixture of both cloud-based and on-premises applications, which are more loosely coupled and can be easily exchanged if needed.

The basic idea is that there should still be a core ERP solution that would cover most important business functions, while other functions will be covered by specialist software solutions that merely extend the core ERP. This concept is similar to the so-called best-of-breed approach to software execution, but it shouldn't be confused with it. While in both cases, applications that make up the whole are relatively loosely connected and quite easily interchangeable, in the case of the latter there is no ERP solution whatsoever. Instead, every business function is covered by a separate software solution.

There is, however, no golden rule as to what business functions should be part of the core ERP, and what should be covered by supplementary solutions. According to Gartner, every company must define their own postmodern ERP strategy, based on company's internal and external needs, operations, and processes. For example, a company may define that the core ERP solution should cover those business processes that must stay behind the firewall, and therefore, choose to leave their core ERP on-premises. At the same time, another company may decide to host the core ERP solution in the cloud and move only a few ERP modules as supplementary solutions to on-premises.

The main benefits that companies will gain from implementing postmodern ERP strategy are speed and flexibility when reacting to unexpected changes in business processes or on the organizational level. With the majority of applications having a relatively loose connection, it is fairly easy to replace or upgrade them whenever necessary. In addition to that, following the examples above, companies can select and combine cloud-based and on-premises solutions that are most suited for their ERP needs. The downside of postmodern ERP is that it will most likely lead to an increased number of software vendors that companies will have to manage, as well as pose additional integration challenges for the central IT.

词汇

生词	音标	释义
coordination	[kəʊˌɔːdɪˈneɪʃn]	*n.* 协作；协调；配合；协调动作的能力
seminal	[ˈsemɪnl]	*adj.* 影响深远的，有重大意义的
entropy	[ˈentrəpɪ]	*n.* 无序状态测法；熵；无序状态
encrypted	[ɪnˈkrɪptɪd]	*v.* 把…加密（或编码）
inversely	[ˌɪnˈvɜː slɪ]	*adv.* 相反地；逆向地；反之
complementary	[ˌkɒmplɪˈmentrɪ]	*adj.* 互补的；补充的；相互补足的
aforementioned	[əˈfɔːmenʃənd]	*adj.* 前面提到的；上述的
acquisition	[ˌækwɪˈzɪʃn]	*n.*（知识、技能等的）获得，得到；（多指贵重的）购得物；购置物；收购的公司；购置的产业；购置；收购
custodianship	[kʌsˈtodɪən, ʃɪp]	*n.* 监督人（保管人）等的身份（或地位）
utility	[juːˈtɪlətɪ]	*n.* 公用事业；实用；效用；有用；实用程序；公用程序 *adj.* 多用途的；多效用的；多功能的
managerial	[ˌmænəˈdʒɪərɪəl]	*adj.* 经理的；管理的
hierarchy	[ˈhaɪərɑːkɪ]	*n.* 等级制度（尤指社会或组织）；统治集团；层次体系
tailored	[ˈteɪləd]	*adj.* 定做的；合身的 *v.* 专门制作；定做
mainframes	[ˈmeɪnfreɪmz]	*n.* 主机；主计算机，大型机
incremental	[ˌɪnkrɪˈmɛnt(ə)l]	*adj.* 增加的；递增的

长难句

There is huge amount of information available to today's manager and this had therefore meant that managers are increasingly relying on management information system to access the exploding information. Management information services helps manager to access relevant, accurate, up-to-date information which is the surer way of making accurate decisions.

今天的管理者可以获得大量信息，因此这意味着管理者越来越依赖管理信息系统来访问大量的信息。管理信息服务可以帮助管理者访问相关的、准确的、最新的信息，这是做出准确决策的可靠方法。

While Material Requirements Planning (MRP) allows for the coordination of raw materials purchasing, Manufacturing Resource Planning Ⅱ (MRP Ⅱ) facilitates the development of a detailed production schedule that accounts for machine and labor capacity, scheduling the production runs according to the arrival of materials.

虽然 MRP 可以协调原材料的采购，但 MRP Ⅱ 可以帮助制订详细的生产计划，该计划应考虑设备物料需求计划和劳动能力，并根据物料的运达制造资源计划安排生产运行。

Reference

[1] David T. Bourgeois. Information systems for business and beyond [M]. America: The Saylo Academy, 2014.
[2] Kroenke D M. What is management information systems? [M]. America: Mays Business School, 2015.
[3] Laudon J P. Leveraging people processes and technology [M]. America: Saunders College of Business, Rochester Institute of Technology, 2017.
[4] Lucey T. Management information systems [M]. England: Cengage Learning EMEA, 2004.
[5] Laudon K C. Management information systems [M]. 11 ed. America: Managing the Digital Firm, Prentice Hall/CourseSmart, 2009.
[6] Boykin G. The history of management information systems [M]. America: Bizfluent retriered. 2018.
[7] Pigni F. A short overview is found in: Luciano Floridi. Information—A Very Short Introduction [M]. England: Oxford University Press, 2010.
[8] Piccolo G. Information systems for managers: With cases [J]. Prospect Press, 2018, 28.
[9] O'Hara M, Watson R, Cavan B. Managing the three levels of change [J]. Information Systems Management, 2018, 16 (3): 64.
[10] Vladimir Z. Information system [M]. England: Britannica, 2016.
[11] D'Atri A, De Marco M, Casalino N. Interdisciplinary aspects of information systems studies [M]. Germany: Physica-Verlag, Springer, 2018.
[12] Jessup L M, Joseph S V. Information systems today [M]. 3rd ed. Glossary: Pearson Publishing, 2018.
[13] Bulgacs S. The first phase of creating a standardised international innovative technological implementation framework/Software application [J]. International Journal of Business and Systems Research, 2018, 7 (3): 250.
[14] Kroenke D M. Experiencing MIS [M]. America: Prentice-Hall, Upper Saddle River, NJ, 2008.
[15] O'Brien J A. Introduction to information systems: essentials for the e-business enterprise [M]. Boston: McGraw-Hill, 2013.
[16] Alter S. 18 Reasons Why IT-Reliant work systems should replace 'The IT Artifact' as the core subject matter of the IS field [J]. Communications of the Association for Information Systems, 2003, 12 (23): 365-394.

[17] Alter S. Work system theory: Overview of core concepts, extensions, and challenges for the future [J]. Journal of the Association for Information Systems, 2013, 14 (2): 72-121.

[18] Alter S. The Work System Method: Connecting People, Processes, and IT for Business Results [M]. Canada: Works System Press, CA, 2006.

[19] Beynon D P. Business information systems [M]. England: Palgrave, Basingstoke, 2009.

[20] Almajali D. Antecedents of ERP systems implementation success: a study on Jordanian healthcare sector [J]. Journal of Enterprise Information Management, 2016, 29 (4): 549-565.

[21] Radovilsky Z, Bidgoli H. The Internet Encyclopedia, Volume 1 [M]. America: John Wiley & Sons, Inc. 2004.

[22] Rubina A, Paula K, Alta van der M. Acceptance of enterprise resource planning systems by small manufacturing Enterprises [C] //In: Proceedings of the 13th International Conference on Enterprise Information Systems, 2009: 229-238.

[23] Shaul L, Tauber D. CSFs along ERP life-cycle in SMEs: a field study [J]. Industrial Management & Data Systems, 2012, 112 (3): 360-384.

[24] Khosrow P M. Emerging trends and challenges in information technology management [M]. America: Idea Group, Inc. 2006.

12 Human Resources Management

Shows in Figure 12 - 1, Robert Owen raised the demand for ten - hour day in year 1810 and instituted it in his New Lanark cotton mills. By 1817 he had formulated the goal of the 8-hour day and coined slogan 8 hours labor, 8 hours recreation, 8 hours full rest.

Figure 12-1 Father of personnel management robert owen
(14 May 1771—17 November 1858)

12.1 Human Resources Management

Human resource management (HRM, or simply HR) is a function in organizations designed to maximize employee performance in service of their employer's strategic objectives (as shown in Figure 12-2). (HRM in changing organizational contexts, 2009). HR is primarily concerned with how people are managed within organizations, focusing on policies and systems (Human resource management: A critical approach, 2009). HR departments and units in organizations are typically responsible for a number of activities, including employee recruitment, training and development, performance appraisal, and rewarding. HR is also concerned with industrial relations, that is, the balancing of organizational practices with regulations arising from collective bargaining and governmental laws.

HR is a product of the human relations movement of the early 20th century when researchers

Figure 12-2 Human resource management

began documenting ways of creating business value through the strategic management of the workforce. The function was initially dominated by transactional work, such as payroll and benefits administration, but due to globalization, company consolidation, technological advancement, and further research, HR now focuses on strategic initiatives like mergers and acquisitions, talent management, succession planning, industrial and labor relations, and diversity and inclusion.

(1) **Antecedent theoretical developments.**

The human resources field began to take shape in 18th century Europe. It built on a simple idea by Robert Owen (1771—1858) and Charles Babbage (1791—1871) during the industrial revolution. These men concluded that people were crucial to the success of an organization. They expressed the thought that the well-being of employees led to perfect work; without healthy workers, the organization would not survive.

HR emerged as a specific field in the early 20th century, influenced by Frederick Winslow Taylor (1856—1915). Taylor explored what he termed "scientific management" (sometimes referred to as "Taylorism"), striving to improve economic efficiency in manufacturing jobs. He eventually focused on one of the principal inputs into the manufacturing process-labor-sparking inquiry into workforce productivity.

Meanwhile, in England, C S Myers, inspired by unexpected problems among soldiers which had alarmed generals and politicians in the First World War of 1914—1918, co-founded the National Institute of Industrial Psychology (NIIP) in 1921. In doing so, he set seeds for the human relations movement. This movement, on both sides of the Atlantic, built on the research of Elton Mayo (1880—1949) and others to document through the Hawthorne studies (1924—1932) and other studies how stimuli, unrelated to financial compensation and working conditions, could yield more productive workers. Work by Abraham Maslow (1908—1970), Kurt Lewin (1890—1947), Max Weber (1864—1920), Frederick Herzberg (1923—2000), and David McClelland (1917—1998), forming the basis for studies in industrial and organizational psychology, organizational behavior and organizational theory, was interpreted in such a way as to further claims of legitimacy for an applied discipline.

(2) **Birth and development of the discipline.**

By the time enough theoretical evidence existed to make a business case for strategic workforce management, changes in the business landscape and in public policy had transformed the employer-employee relationship, and the discipline became formalized as "industrial and labor relations". In 1913 one of the oldest known professional HR associations—the Chartered Institute of Personnel and Development (CIPD) —started in England as the Welfare Workers' Association; it changed its name a decade later to the Institute of Industrial Welfare Workers, and again the next decade to Institute of Labor Management before settling upon its current name in 2000. In 1920, James R. Angell delivered an address to a conference on personnel research in Washington detailing the need for personnel research. This preceded and led to the organization of the personnel research federation. In 1922, the first volume of the journal of personnel research was published, a joint initiative between the national research council and the engineering foundation. Likewise in the United States, the world's first institution of higher education dedicated to workplace studies—the School of Industrial and Labor Relations—formed at Cornell University in 1945. In 1948 what would later become the largest professional HR association—the Society for Human Resource Management (SHRM) —formed as the American Society for Personnel Administration (ASPA).

In the Soviet Union, meanwhile, Stalin's use of patronage exercised through the "HR Department" equivalent in the Bolshevik Party, its Orgburo, demonstrated the effectiveness and influence of human-resource policies and practices, and Stalin himself acknowledged the importance of the human resource, such as in his mass deployment of it in the Gulag system.

During the latter half of the 20th century, union membership declined significantly, while workforce management continued to expand its influence within organizations. In the US, the phrase "industrial and labor relations" came into use to refer specifically to issues concerning collective representation, and many companies began referring to the proto-HR profession as "personnel administration". Many current HR practices originated with the needs of companies in the 1950s to develop and retain talent.

In the late 20th century, advances in transportation and communications greatly facilitated workforce mobility and collaboration. Corporations began viewing employees as assets. "Human resources management" consequently, became the dominant term for the function—the ASPA even changing its name to the Society for Human Resource Management (SHRM) in 1998.

"Human capital management" (HCM) is sometimes used synonymously with "HR", although "human capital" typically refers to a more narrow view of human resources, i. e. the knowledge the individuals embody and can contribute to an organization. Likewise, other terms sometimes used to describe the field include "organizational management", "manpower management", "talent management", "personnel management", and simply "people management".

12. 1. 1 Human Resources Management Overview

Human Resource Management (HRM or HR) is the strategic approach to the effective management of people in a company or organization such that they help their business gain a competitive advantage. Figure 12 – 3 shows the process of HRM. It is designed to maximize employee performance in service of an employer' s strategic objectives[1] . Human resource management is primarily concerned with the management of people within organizations, focusing on policies and systems. HR departments are responsible for overseeing employee-benefits design, employee recruitment, training and development, performance appraisal, and reward management, such as managing pay and benefit systems. HR also concerns itself with organizational change and industrial relations, or the balancing of organizational practices with requirements arising from collective bargaining and governmental laws. As per Armstrong (1997), Human Resource Management can be characterized as " a vital way to deal with obtaining, creating, overseeing, rousing and picking up the dedication of the association' s distinct advantage——the individuals who work in and for it"[1].

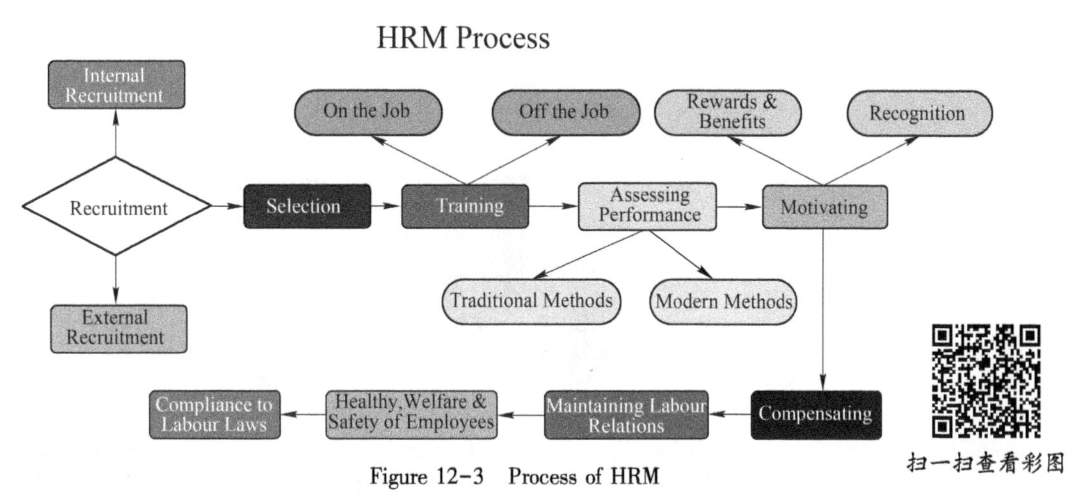

Figure 12-3 Process of HRM

The overall purpose of human resources (HR) is to ensure that the organization is able to achieve success through people. HR professionals manage the human capital of an organization and focus on implementing policies and processes. They can specialize in finding, recruiting, training, and developing employees, as well as maintaining employee relations or benefits. Training and development professionals ensure that employees are trained and have continuous development. This is done through training programs, performance evaluations, and reward programs. Employee relations deals with the concerns of employees when policies are broken, such as cases involving harassment or discrimination. Managing employee benefits includes developing compensation structures, parental leave programs, discounts, and other benefits for employees. On the other side of the field are HR generalists or business partners. These HR professionals could work in all areas or be labor relations representatives working with unionized employees. [2]

HR is a product of the human relations movement of the early 20th Century, when researchers began documenting ways of creating business value through the strategic management of the workforce[3]. It was initially dominated by transactional work, such as payroll and benefits administration, but due to globalization, company consolidation, technological advances, and further research, HR as of 2015 focuses on strategic initiatives like mergers and acquisitions, talent management, succession planning, industrial and labor relations, and diversity and inclusion. In the current global work environment, most companies focus on lowering employee turnover and on retaining the talent and knowledge held by their workforce. New hiring not only entails a high cost but also increases the risk of a new employee not being able to adequately replace the position of the previous employee. HR departments strive to offer benefits that will appeal to workers, thus reducing the risk of losing employee commitment and psychological ownership.

12.1.2 Strategic Human Resource Management

Figure 12-4 shows strategic human resource management.

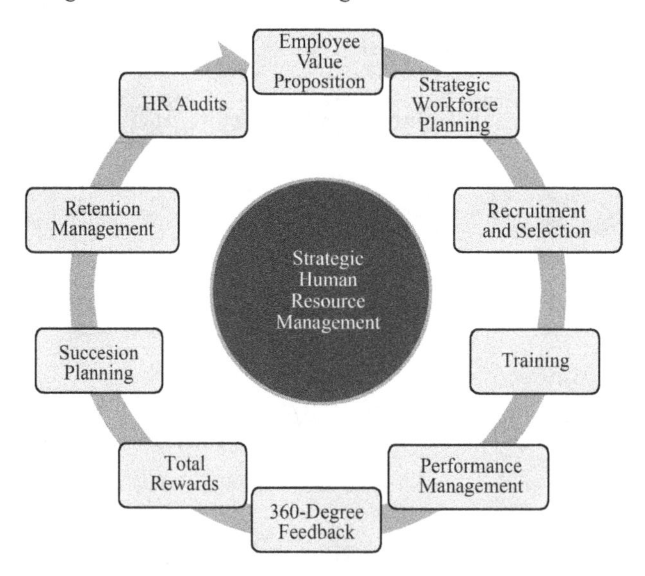

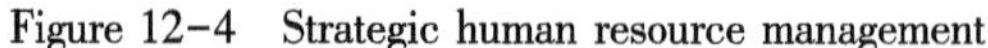
Figure 12-4 Strategic human resource management

12.1.2.1 Strategic Human Resource Management

Strategic human resource management is "critical importance of human resources to strategy, organizational capability to adapt to change and the goals of the organization". In other words, this is a strategy that intends to adapt the goals of an organization and is built off of other theories such as the contingency theory as well as institutional theory which fit under the umbrella of organizational theory. These theories look at the universalize, contingency and configuration perspectives to see the effect of human resource practices in organizations. The universalize perspective says that there are better human resource practices than others and those should be

adopted within organization while contingency says that human resource practices need to align with other organization practices or the organizations mission, and configuration perspective is based on how to combine multiple aspects of human resource practices with effectiveness or performance. This can also be viewed as how human resource practices fit vertically or horizontally in an organization. This theory also involves looking at the value of human capital as well as social capital both in and outside of organizations and how this affects human resource practices. Human capital being knowledge and skills of individuals working for the organization and social capital is based on the character and value of relationships in and out of the organization. "Colbert suggests that SHRM should focus on the interactions and processes of the organization's social system—the intentions, choices and actions of people in the system and on HR systems as a coherent whole."

Strategic human resource management is the connection between a company's human resources and its strategies, objectives, and goals. The aim of strategic human resource management is to: (1) Advance flexibility innovation, and competitive advantage. (2) Develop a fit for purpose organizational culture. (3) Improve business performance.

In order for strategic human resource management to be effective, human resources (HR) must play a vital role as a strategic partner when company policies are created and implemented. Strategic HR can be demonstrated throughout different activities, such as hiring, training, and rewarding employees.

Strategic HR involves looking at ways that human resources can make a direct impact on a company's growth. HR personnel need to adopt a strategic approach to developing and retaining employees to meet the needs of the company's long-term plans.

HR issues can be a difficult hurdle to cross for many companies, there are all kinds of different components that can confuse business owners and cause them to make ineffective decisions that slow down the operations for their employees as well as their business. HR departments that practice strategic human resource management do not work independently within a silo, they interact with other departments within an organization in order to understand their goals and then create strategies that align with those objectives, as well as those of the organization. As a result, the goals of a human resource department reflect and support the goals of the rest of the organization. Strategic HRM is seen as a partner in organizational success, as opposed to a necessity for legal compliance or compensation. Strategic HRM utilizes the talent and opportunity within the human resources department to make other departments stronger and more effective.

12. 1. 2. 2 Why is Strategic Human Resource Management Important

Companies are more likely to be successful when all teams are working towards the same objectives. Strategic HR carries out analysis of employees and determines the actions required to increase their value to the company. Strategic human resource management also uses the results of this analysis to develop HR techniques to address employee weaknesses.

The following are benefits of strategic human resource management:

(1) Increased job satisfaction.
(2) Better work culture.
(3) Improved rates of customer satisfaction.
(4) Efficient resource management.
(5) Proactive approach to managing employees.
(6) Boost productivity.

12. 1. 2. 3 Seven Steps to Strategic Human Resource Management

Strategic human resource management is key for the retention and development of quality staff. It's likely that employees will feel valued and want to stay with a company that places a premium on employee retention and engagement. Before implementing strategic human resource management, you will need to create a strategic HR planning process using the steps below:

(1) Develop a thorough understanding of company's objectives;
(2) Evaluate HR capability;
(3) Analyze current HR capacity in light of your goals;
(4) Estimate company's future HR requirements;
(5) Determine the tools required for employees to complete the job;
(6) Implement the human resource management strategy;
(7) Evaluation and corrective action.

Strategic human resource management is important for every company. Company doesn't need to employ a specific number of employees before start to consider implementing strategic human resource management principles. In fact, if we have a plan to grow business, we should be thinking about linking this growth to strategic human resource management. Some companies outsource this part of their business because they don't have an in-house HR function. Strategic human resource services provide full-service HR functions including developing a human resource management strategy. Strategic HR services help to take away the burden of both operational and strategic management to facilitate the growth of your business.

12. 1. 3 Human Resource Management Environment

External Environment (External Environment of HRM) refers to various factors that can influence the HRM environment outside the enterprise system. Generally speaking, the external environment of HUMAN resource management can be analyzed from the aspects of political environment, economic factors, cultural factors, and competitors. Because these influencing factors are outside the scope of enterprises, enterprises cannot directly control and influence them. In most cases, they can only take corresponding measures according to the status and changes of the external environment.

Internal environment (internal environment of HRM) refers to various factors that can influence human resource management activities within the enterprise system. Since human resource is one of the essential elements for any enterprise to maintain normal activities, and human resource management also runs through all aspects of enterprise production and operation, in this sense, all the factors that constitute an enterprise are the internal environment of human resource management. However, the internal environment of human resource management is usually analyzed from the aspects of enterprise development strategy, organizational structure, enterprise life cycle and enterprise culture. Different from the external environment, various factors of the internal environment are within the scope of the enterprise, so the enterprise can directly influence them.

12.1.4 Hiring

Recruitment refers to the overall process of identifying, attracting, screening, shortlisting, and interviewing, suitable candidates for jobs (either permanent or temporary) within an organization. Recruitment can also refer to processes involved in choosing individuals for unpaid roles. Managers, human resource generalists and recruitment specialists may be tasked with carrying out recruitment, but in some cases public-sector employment agencies, commercial recruitment agencies, or specialist search consultancies are used to undertake parts of the process. Internet-based technologies which support all aspects of recruitment have become widespread, including the use of Artificial Intelligence (AI).

12.2 Human Resource Planning (HRP)

Human Resource Planning (**HRP**, as shown in Figure 12-5) is the process of forecasting the future human resource requirements of the organization and determining as to how the existing human resource capacity of the organization can be utilized to fulfill these requirements. It, thus, focuses on the basic economic concept of demand and supply in context to the human resource capacity of the organization. Figure 12-6 shows the steps of human resource planning.

扫一扫查看彩图

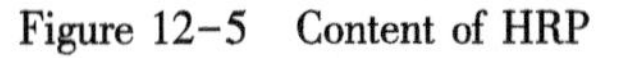
Figure 12-5 Content of HRP

It is the HRP process which helps the management of the organization in meeting the future demand of human resource in the organization with the supply of the appropriate people in appropriate numbers at the appropriate time and place. Further, it is only after proper analysis of

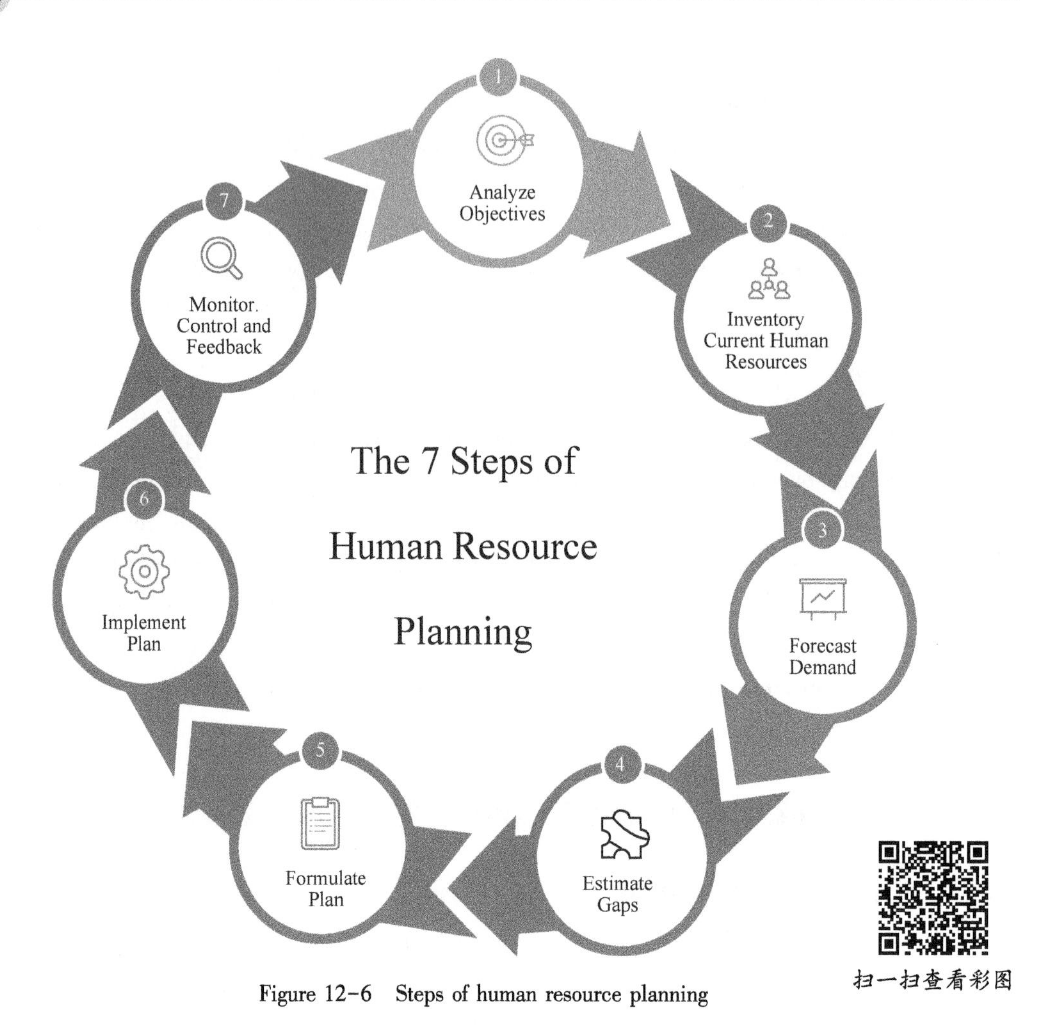

Figure 12-6 Steps of human resource planning

the HR requirements can the process of recruitment and selection be initiated by the management. Also, HRP is essential in successfully achieving the strategies and objectives of organization. In fact, with the element of strategies and long-term objectives of the organization being widely associated with human resource planning these days, HR planning has now become strategic HR planning.

Though, HR planning may sound quite simple a process of managing the numbers in terms of human resource requirement of the organization, yet the actual activity may involve the HR manager to face many roadblocks owing to the effect of the current workforce in the organization, pressure to meet the business objectives and prevailing workforce market condition. HR planning, thus, help the organization in many ways as follows:

(1) HR managers are in a stage of anticipating the workforce requirements rather than getting surprised by the change of events.

(2) Prevent the business from falling into the trap of shifting workforce market, a common concern among all industries and sectors.

(3) Work proactively as the expansion in the workforce market is not always in conjunction

with the workforce requirement of the organization in terms of professional experience, talent needs, skills, etc.

(4) Organizations in growth phase may face the challenge of meeting the need for critical set of skills, competencies, and talent to meet their strategic objectives so they can stand well-prepared to meet the HR needs.

(5) Considering the organizational goals, HR planning allows the identification, selection and development of required talent or competency within the organization.

It is, therefore, suitable on the part of the organization to opt for HR planning to prevent any unnecessary hurdles in its workforce needs. An HR consulting firm can provide the organization with a comprehensive HR assessment and planning to meet its future requirements in the most cost-effective and timely manner.

An HR planning process simply involves the following four broad steps:

(1) Current HR supply: Assessment of the current human resource availability in the organization is the foremost step in HR planning. It includes a comprehensive study of the human resource strength of the organization in terms of numbers, skills, talents, competencies, qualifications, experience, age, tenures, performance ratings, designations, grades, compensations, benefits, etc. At this stage, the consultants may conduct extensive interviews with the managers to understand the critical HR issues they face and workforce capabilities they consider basic or crucial for various business processes.

(2) Future HR demand: Analysis of the future workforce requirements of the business is the second step in HR planning. All the known HR variables like attrition, lay-offs, foreseeable vacancies, retirements, promotions, pre-set transfers, etc. are taken into consideration while determining future HR demand. Further, certain unknown workforce variables like competitive factors, resignations, abrupt transfers, or dismissals are also included in the scope of analysis.

(3) Demand forecast: Next step is to match the current supply with the future demand of HR and create a demand forecast. Here, it is also essential to understand the business strategy and objectives in the long run so that the workforce demand forecast is such that it is aligned to the organizational goals.

(4) HR sourcing strategy and implementation: After reviewing the gaps in the HR supply and demand, the HR consulting firm develops plans to meet these gaps as per the demand forecast created by them. This may include conducting communication programs with employees, relocation, talent acquisition, recruitment and outsourcing, talent management, training and coaching, and revision of policies. The plans are, then, implemented taking into confidence the mangers so as to make the process of execution smooth and efficient. Here, it is important to note that all the regulatory and legal compliances are being followed by the consultants to prevent any untoward situation coming from the employees.

Hence, a properly conducted process of HR planning by an HR consulting firm helps the organization in meeting its goals and objectives in timely manner with the right HR strength in action. Administration and operations used to be the two role areas of HR. The strategic planning component came into play as a result of companies recognizing the need to consider HR needs in goals and strategies. HR directors commonly sit on company executive teams because of the HR planning function. Numbers and types of employees and the evolution of compensation systems are among elements in the planning role [4]. Various factors affecting Human Resource planning Organizational Structure, Growth, Business Location, Demographic changes, environmental uncertainties, expansion etc. Additionally, this area encompasses the realm of talent management.

12.2.1 Human Resource Demand and Supply Forecast

12.2.1.1 HR Supply Forecasting Introduction

(1) Human resource supply forecasting is the process of estimating availability of human resource followed after demand for testing of human resource.

(2) Supply forecasting means to make an estimation of supply of human resources taking into consideration the analysis of current human resources inventory and future availability.

(3) For forecasting supply of human resource we need to consider internal and external supply.

(4) Ecall of laid off employees, etc.

(5) Source of external supply of human resource is availability of Laboure force in the market and new recruitment.

In microeconomics, **supply and demand** is an economic model of price determination in a market. It postulates that, holding all else equal, in a competitive market, the unit price for a particular good, or other traded item such as labor or liquid financial assets, will vary until it settles at a point where the quantity demanded (at the current price) will equal the quantity supplied (at the current price), resulting in an economic equilibriumfor price and quantity transacted.

The first step of forecasting the future supply of human resource is to obtain the data and information about the present human resources inventory.

Existing inventory: the data relating to present human resource inventory in terms of human resources components, number, designation wise and department wise should be obtained.

Principal dimension of human resources inventory are:

(1) Head counts regarding total, department-wise, sex-wise, designation-wise, pay roll - wise.

(2) Job family inventory: it includes number and categories of employees of each job family such as all jobs relating to the same categories as clerk, cashier, typist and steno, each sub job family such as all jobs having common jobs characteristics (skill, qualification, similar operations).

(3) Age inventory: It includes age-wise number and category of employees. It indicates age - wise imbalance in present inventory which can be correlated in future selections and promotion.

12.2.1.2 Estimating The Net Human Resource Requirements

Net human resource requirements in terms of numbers and components are to be determined in relation to the overall human resources requirement (demand forecast) for future date and supply for a future date and supply forecast for that date. The difference between overall human requirements and future supply of human resources is to be found out. This difference is the net human resource requirements.

12.2.1.3 Action Plan For Redeployment, Redundancy/Retrenchment

If future surplus is estimated, the organization has to plan for redeployment, redundancy. If surplus estimated in some jobs/departments, employees can be redeployed in other jobs/departments, where the deficit of employees is estimated. The organization should also plan for training or reorientation before redeployment of employees. If the deficit is not estimated in any job/department and surplus is estimated for the entire organization, the organization in consultation with the trade unions has to plans for redundancy or retrenchment.

12.2.1.4 Redeployment Programmes

The redeployment programs are as follow:

Outplacement: outplacement programs also intended to provide career guidance for displaced employees. This program covers retraining the prospective displaced employees who can outplacement programs also intended to provide career guidance for displaced employees. This program covers retraining the prospective displaced employees who can deployed elsewhere in the organization, helping in resume writing, interview techniques job searching .

12.2.1.5 Redundancy/Retrenchment Programmes

The redundancy /retrenchment programs include:

(1) Reduced work hours: under this technique, each worker works less hours and receive less pay, so that the job is saved.

(2) Work sharing: some organization offer employees the opportunity to share jobs pr two employees work half time each. This technique solves the problem of retrenchment in the short run.

(3) Layoffs: layoffs can be temporary or permanent. Temporary layoff is due to the slackness in business, machinery breakage, power failure. Workers is called back as soon as work resumes to the normal position. Permanent layoff is due to liquidation of the company. Proper human resource planning and leveling the workforce at proper level can help to reduce this effect.

(4) Leave of absence without pay: This technique helps the company to cut the labor cost and the employee to pursue his self-interest. This technique also helps the company to plan for eliminating the unnecessary job in a phased manner.

(5) Voluntary retirement /early retirement: Another issue is early retirement. Government of India introduced voluntary retirement scheme (VRS) under the caption Golden Handshake, in order to solve the problem of overstaffing in the public sector. Management provides cash reward to those employees who opt for VRS in addition to normal retirement benefits. Hence, this is also

called golden handshake. This technique solves the problem of excessive supply of future inventory over the demand of human resource.

(6) Attrition: is the process whereby an incumbents leave their job for various reason; those jobs will be kept vacant or unfilled. Attrition, hiring freezes on employment can be implemented organization -wise or job wise.

(7) Compulsory retirement /iron handshake: under this program, the HR manager, with the help of the line manager, identifies surplus employees and discharges them from the services. management do not provide any cash or non- cash benefits to the employees other than normal retirement benefits at the time of discharging or firing.

12.2.1.6 Forecast Future from all the Sources

If deficit is estimated in any department and in the entire organization, management has to forecast the future supply of human resources from various source like internal sources, comparable organization, education and training institutes, employment exchanges and labor market.

12.2.2 Company Organizational Structure Design

Organizational structure (as shown in Figure 12-7) is used to develop how groups and individuals are arranged or departmentalized to help meet an organization's goals. It defines a reporting structure, jobs, compensation, and responsibilities for each role. Designing an organizational structure

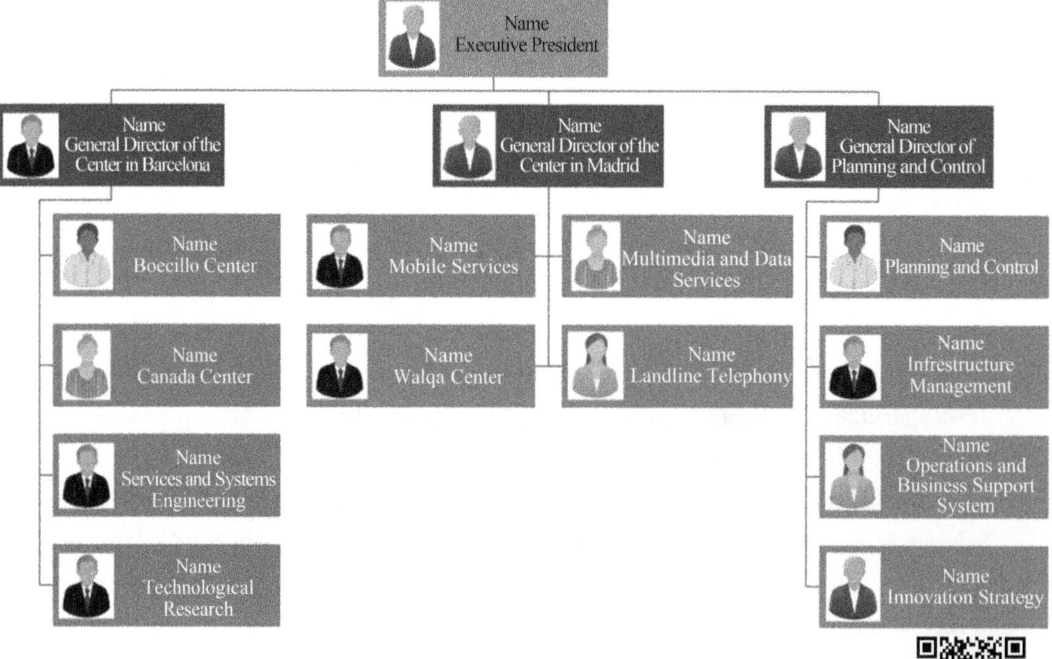

Figure 12-7 Company organizational structure chart

扫一扫查看彩图

requires consideration of an organization's values, financial and business goals. It should allow for growth for the organization and the ability to add additional jobs or departments.

(1) Define business units or departments. Each business unit should have similar goals and responsibilities that can be overseen and directed by one or several managers. The business units or departments will then align to assist in creating an appropriate organizational structure. Depending on which type of organizational structure is used, departments may align laterally with other departments, or one may oversee another.

(2) Determine which type of organizational structure best fits your business needs. The several types of organizational structure ensure an organization can successfully function with its reporting structure, expand if necessary and successfully meet its goals. For example, if your organization is small, it may simply require the organizational structure be broken into departments, such as production, human resources, and finance. Your organization's business type, units and how it operates will determine which type of organizational structure to choose.

(3) Define the executive and management teams. Executives and managers are responsible for ensuring each business unit meets the organization's goals. This may include one or several top executives to oversee the entire organization and managers to direct each business unit within the organizational structure. The organization may require one supervisor to oversee all operations, or several supervisors to direct each business unit, ultimately reporting to a top executive or owner.

(4) Establish performance metrics and compensation. When the organizational structure is determined, job descriptions can be clearly defined and where each job fits in the hierarchy. Each job description should reflect the competencies required to do the job and the expectations of each job to meet the organization's goals. After each job within the structure is defined, compensation should be defined based on the responsibilities of each job.

A normal corporate structure consists of various departments that contribute to the company's overall mission and goals. Common departments include Marketing, Finance, Operations management, Human Resource, and IT. These five divisions represent the major departments within a publicly traded company, though there are often smaller departments within autonomous firms. There is typically a CEO, and Board of Directors not usually composed of the directors of each department. There are also company presidents, vice presidents, and CFOs. There is a great diversity in corporate forms as enterprises may range from single company to multi-corporate conglomerate. The four main corporate structures are Functional, Divisional, Geographic, and the Matrix. Realistically, most corporations tend to have a "hybrid" structure, which is a combination of different models with one dominant strategy.

词汇

生词	音标	释义
recruitment	[rɪˈkruːtmənt]	*n.* 招收，招聘；（自然种群）增长；募集（反应、现象）
transactional	[trænˈzækʃənl]	*adj.* 交易型的；事务性的；事务处理的

consolidatin	[kənˌsɒlɪˈdeɪʃən]	*n.* 巩固；合并；团结
implement	[ˈɪmplɪmənt]	*n.* 工具，器具；手段 *vt.* 实施，执行；实现，使生效
specialize	[ˈspeʃəlaɪz]	*vi.* 专门从事；详细说明；特化 *vt.* 使专门化；使适应特殊情况；详细说明
discrimination	[dɪsˌkrɪmɪˈneɪʃn]	*n.* 歧视；区别，辨别；识别力
critical	[ˈkrɪtɪkl]	*adj.* 鉴定的；［核］临界的；批评的，爱挑剔的；危险的；决定性的；评论的
contingency	[kənˈtɪndʒənsi]	*n.* 可能发生的事；不能确定的事件；应急措施；应急储备；应急开支；可能性；意外；胜诉酬金 *adj.* 应变的
configuration	[kənˌfɪgʊˈreɪʃən]	*n.* 配置；结构；外形
perspective	[pəˈspektɪv]	*n.* 观点；远景；透视图 *adj.* 透视的
coherent	[koʊˈhɪərənt]	*adj.* 连贯的，一致的；明了的；清晰的；凝聚性的；互相耦合的；粘在一起的；共格的
microeconomics	[ˈmaɪkrəuˌiːkəˈnɒmɪks]	*n.* 微观经济学
corporate	[ˈkɔːprɪt]	*adj.* 法人的；共同的，全体的；社团的；公司的；企业的
appraisal	[əˈpreɪzl]	*n.* 评价；估价（尤指估价财产，以便征税）
payroll	[ˈpeɪˌrəʊl]	*n.* 工资单；在册职工人数；工资名单；工资
institutional	[ˌɪnstɪˈtuːʃənl]	*adj.* 制度的；制度上的学会的；由来已久的；习以为常的公共机构的；慈善机构的
proactively	[ˌprəʊˈæktɪvlɪ]	*adv.* 前摄地；主动地
cost-effective	[ˌkɒst ɪˈfektɪv]	*adj.* 划算的；成本效益好的
roadblocks	[ˈrəʊdblɒk]	*n.* 路障；障碍物 *vi.* 设置路障

长难句

The universalize perspective says that there are better human resource practices than others and those should be adopted within organization while contingency says that human resource practices need to align with other organization practices or the organizations mission, and configuration perspective is based on how to combine multiple aspects of human resource practices with effectiveness or performance.

普遍化观点认为，总有更好的人力资源实践方法，应在组织内部采用，而另一种相依关联的观点认为，人力资源的实践需要与其他组织实践或组织任务保持一致，并且配置视角是基于如何将人力资源实践的多个方面与有效性或绩效结合起来。

It postulates that, holding all else equal, in a competitive market, the unit price for a particular good, or other traded item such as labor or liquid financial assets, will vary until it settles at a point where the quantity demanded (at the current price) will equal the quantity supplied (at the current price), resulting in an economic equilibrium for price and quantity transacted.

它假定，如果一切平等，在竞争激烈的市场中，一个特定商品的单价，或其他交易商品比如劳动力或流动金融资产，会有所不同，直到平稳在一个点，即需求量（在当前价格）将等于供给量（在目前的价格），这就是形成价格与交易数量的经济均衡。

Reference

[1] Akingbola K. A model of strategic nonprofit human resource management [J]. Voluntas, 2012, 24 (1): 214-240.

[2] Sulich A. Mathematical models and non-mathematical methods in recruitment and selection processes [R]. Mekon, 17th International Conference, 2015.

[3] Roman T, Stephen B, Rob Q. Corporation law in Australia [M]. Australia: Federation Press, 2002.

[4] Lucey. Designing an effective organization structure [M]. Boston: Bain & Company organizational toolkit and Bridgespan analysis, 2014.

13 Intelligent Manufacturing

13.1 Intelligent Manufacturing Technology

13.1.1 The Development, Connotation, and Characteristics of Intelligent Manufacturing Technology

13.1.1.1 The Development

Countries around the world are actively engaging in the new industrial revolution. The United States has launched the advanced manufacturing partnership, germany has developed the strategic initiative Industry 4.0, and the United Kingdom has put forward the UK Industry 2050 strategy. In addition, France has unveiled the new industrial france program, Japan has a Society 5.0 strategy, and Korea has started the Manufacturing Innovation 3.0 program. The development of intelligent manufacturing is regarded as a key measure to establish competitive advantages for the manufacturing industry of major countries around the world. The made in China 2025 plan, formerly known as China Manufacturing 2025, has specifically set the promotion of intelligent manufacturing as its main direction, with a focus on the in-depth integration of new-generation information technology within the manufacturing industry.

Since the beginning of the 21st century, new-generation information technology has shown explosive growth. The broad application of digital, networked, and intelligent technologies in the manufacturing industry and the continuous development of integrated manufacturing innovations have been the main driving forces of the new industrial revolution. **In particular, new-generation intelligent manufacturing, which serves as the core technology of the current industrial revolution, incorporates major and profound changes in the development philosophy, manufacturing modes, and other aspects of the manufacturing industry.** Intelligent manufacturing is now reshaping the development paths, technical systems, and industrial forms of the manufacturing industry, and is thereby pushing the global manufacturing industry into a new stage of development. Industry 4.0 is a project in the high-tech strategy of the German government that promotes the computerization of traditional industries such as manufacturing. The goal is the intelligent factory (smart factory) that is characterized by adaptability, resource efficiency, and ergonomics, as well as the integration of customers and business partners in business and value processes. Its technological foundation consists of cyber-physical systems and the internet of things[1].

13. 1. 1. 2 Three Basic Paradigms of Intelligent Manufacturing

Intelligent manufacturing is a general concept that covers a wide range of specific topics. New-generation intelligent manufacturing represents an in-depth integration of new-generation artificial intelligence (AI) technology and advanced manufacturing technology. It runs through every link in the full life cycle of design, production, product, and service. The concept also relates to the optimization and integration of corresponding systems; it aims to continuously raise enterprises' product quality, performance, and service levels while reducing resources consumption, thus promoting the innovative, green, coordinated, open, and shared development of the manufacturing industry.

For decades, intelligentization for manufacturing has involved many different paradigms as it continues to develop in practice. These paradigms include lean production, flexible manufacturing, concurrent engineering, agile manufacturing, digital manufacturing, computer-integrated manufacturing, networked manufacturing, cloud manufacturing, intelligent manufacturing, and more. All of these paradigms have played an active role in guiding technology upgrading in the manufacturing industry. However, there are too many paradigms to form a unified intelligent manufacturing technology roadmap; this lack of unity causes enterprises to experience many perplexities in their practice of pushing forward intelligent upgrading. Considering the continuously emerging new technologies, new ideas, and new modes of intelligent manufacturing, we consider it necessary to summarize the basic paradigms of intelligent manufacturing.

Intelligent manufacturing has developed in parallel with the progress of informatization. There are three stages in the development of informatization worldwide:

From the middle of the 20th century to the mid 1990s, informatization was in a digital stage with computing, communications, and control applications as the main features.

Starting in the mid 1990s, the Internet came into large-scale popularization and application, and informatization entered a networked stage with the interconnection of all things as its main characteristic.

At present, on the basis of cluster breakthroughs in and integrated applications of big data, cloud computing, the mobile Internet, and the Industrial Internet, strategic breakthroughs have been made in AI; as a result, informatization has entered an intelligent stage, with new-generation AI technology as its main feature.

Taking the various intelligent manufacturing-related paradigms into account and considering the characteristics of the integration of information technology and the manufacturing industry through different stages, it is possible to generalize three basic paradigms of intelligent manufacturing: digital manufacturing, digital-networked manufacturing, and new-generation intelligent manufacturing. Figure 13-1 shows the evolution of three basic paradigms of intelligent manufacturing.

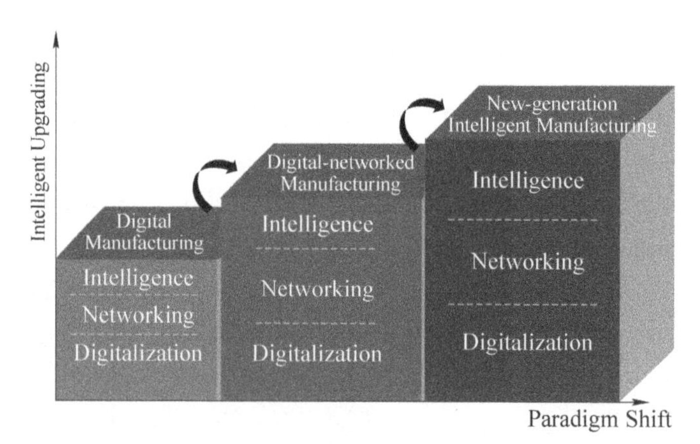

Figure 13-1 The evolution of three basic paradigms of intelligent manufacturing

13.1.1.3 Digital Manufacturing

Digital manufacturing is the first basic paradigm of intelligent manufacturing; it may also be referred to as first-generation intelligent manufacturing.

The concept of intelligent manufacturing first appeared in the 1980s. Because the first-generation AI technology that was in application at that time could hardly solve specific engineering problems, first-generation intelligent manufacturing was essentially digital manufacturing.

Starting in the second half of the 20th century, as demand for technological progress in the manufacturing sector became increasingly urgent, digital information technologies were widely applied in the manufacturing industry, driving forward revolutionary changes in the industry. Against a background of the integration of digital technology with manufacturing technology, digital manufacturing undertook the digital description, analysis, decision-making, and control of product information, process information, and resources information; in this way, digital manufacturing remarkably shortened the time required for designing and manufacturing products to meet specific customer requirements.

The key features of digital manufacturing are as follows:

Digital technology is widely used in products, forming a "digital generation" of innovative products; digital design, modeling and simulations, and digital equipment information management are widely applied, and production process integration and optimization are achieved.

The point that needs to be clarified here is that digital manufacturing is the foundation of intelligent manufacturing. Digital manufacturing continues to evolve and runs throughout the three basic paradigms and all the development processes of intelligent manufacturing. The digital manufacturing being defined here is the digital manufacturing of the first basic paradigm, which positions digital manufacturing in a relatively narrow sense. On an international level, several types of positioning and theories on digital manufacturing have also been developed in a broad sense. Some of the key technologies in the smart manufacturing movement include big data processing capabilities, industrial connectivity devices and services, and advanced robotics.

13.1.1.4 Digital-Networked Manufacturing

Digital-networked manufacturing is the second basic paradigm of intelligent manufacturing; it may also be referred to as "Internet + manufacturing" or as second-generation intelligent manufacturing.

In the end of the 20th century, Internet technology started to gain popularity. "Internet +" has continuously pushed forward the integrated development of the Internet and the manufacturing industry. The network connects humans, processes, data, and things. Through intra- and inter-enterprise collaborations and the sharing and integration of all kinds of social resources, "Internet +" reshapes the value chain of the manufacturing industry and drives the transformation from digital manufacturing to digital-networked manufacturing.

The main characteristics of digital-networked manufacturing are as follows:

At the product level, digital technology and network technology are widely applied. Products are connected through the network, while collaborative and shared design and R&D are achieved.

At the manufacturing level, horizontal integration, vertical integration, and end-to-end integration are completed, thereby connecting the data flows and information flows of the entire manufacturing system.

At the service level, enterprises and users connect and interact through the network platforms, while enterprises begin to transform from product-centered production to user-centered production.

Both Germany's Industry 4.0 report and General Electric's Industrial Internet report present very informative and well-structured descriptions of the digital-networked manufacturing paradigm and put forward technology roadmaps for digital-networked manufacturing.

13.1.1.5 New-Generation Intelligent Manufacturing

Digital-networked-intelligent manufacturing is the third basic paradigm of intelligent manufacturing; it may also be referred to as new-generation intelligent manufacturing.

Jointly driven by a strong demand for economic and social development, the penetration of the Internet, the emergence of cloud computing and big data, the development of the Internet of Things, and rapid changes in the information environment, there has been an accelerating development of strategic breakthroughs in new-generation AI technologies; these include big data intelligence, human-machine hybrid-augmented intelligence, crowd intelligence, and cross-media intelligence. The in-depth integration of new-generation AI technology and advanced manufacturing technology leads to the formation of new-generation intelligent manufacturing. New-generation intelligent manufacturing will reshape all the processes of the full product cycle, including design, manufacture, and services, as well as the integration of these processes. It will promote the emergence of new technologies, new products, new business forms, and new models, and it will profoundly influence and change the production structure, production modes, lifestyles, and thinking models of humankind. It will ultimately result in a great improvement of social productive forces. New-generation intelligent manufacturing will bring revolutionary changes to the manufacturing industry and will become the main driving force for the future development of the industry.

The three basic paradigms of intelligent manufacturing reflect the inherent development pattern of intelligent manufacturing. On the one hand, the three basic paradigms developed in sequence, each with their own characteristics and key problems to solve; in this way, they embody the characteristics of different developmental stages of advanced information technology and advanced manufacturing technology. On the other hand, the three basic paradigms cannot be technologically separated from each other; rather, they are interconnected and iteratively upgraded, thus showing the integrated development characteristics of intelligent manufacturing. China and other emerging industrial countries must leverage their late-mover advantages and adopt a technology roadmap for the "parallel promotion and integrated development" of the three basic paradigms.

13.1.2 Advanced Manufacturing

Advanced manufacturing is the use of innovative technology to improve products or processes, with the relevant technology being described as "advanced", "innovative", or "cutting edge". Advanced manufacturing industries "increasingly integrate new innovative technologies in both products and processes. The rate of technology adoption and the ability to use that technology to remain competitive and add value to define the advanced manufacturing sector."[2]

World class manufacturing (WCM) "integrates the latest-generation machinery with (process/work) systems to facilitate 'manufacturing' - based business development governed around manufactured[3] products only, duly based over a high accent on Product Substitution or New Product Development."

"Advanced manufacturing centers upon improving the performance of US industry through the innovative application of technologies, processes and methods to product design and production."[4] A 2010 survey of advanced manufacturing definitions by the White House states: "A concise definition of advanced manufacturing offered by some is manufacturing that entails the rapid transfer of science and technology (S&T) into manufacturing products and processes." (PCAST, April 2010.)

13.1.2.1 Product Technologies

Organizations practicing advanced manufacturing make products characterized as:

(1) Products with high levels of design;

(2) Technologically complex;

(3) Innovative;

(4) Reliable, affordable, and available;

(5) Newer, better, more exciting;

(6) Products that solve a variety of problems;

(7) Flexibility.

13.1.2.2 Process Technologies

The manufacturing process technologies described in definitions of advanced manufacturing include[5]:

(1) Computer technologies (e.g., CAD, CAE, CAM) (Paul Fowler, NACFAM; UK Manufacturing Advisory Service Southeast; C. B. Adams, St. Louis, OECD);

(2) High-Performance Computing (HPC) for modeling, simulation, and analysis (Council on Competitiveness);

(3) Rapid prototyping (additive manufacturing);

(4) High Precision technologies (Paul Fowler, NACFAM);

(5) Information technologies (Paul Fowler, NACFAM);

(6) Advanced robotics and other intelligent production systems (US Department of Labor, ETA; C. B. Adams, St. Louis);

(7) Automation (UK Manufacturing Advisory Service Southeast; C. B. Adams, St. Louis);

(8) Control systems to monitor processes (UK Manufacturing Advisory Service Southeast);

(9) Sustainable and green processes and technologies (US Department of Labor);

(10) New industrial platform technologies (e.g., composite materials) (UK Manufacturing Advisory Service Southeast);

(11) Ability to custom manufacture (PCAST; Paul Fowler, NACFAM; Grow Oklahoma Campaign)[6];

(12) Ability to manufacture high or low volume (scalability) (PCAST; Paul Fowler, NACFAM; Grow Oklahoma Campaign);

(13) High rate of manufacturing.

13.2 Digital Twin

13.2.1 Digital Twin Factory

A digital twin is a digital replica of a living or non-living physical entity. Digital twin refers to a digital replica of potential and actual physical assets (physical twin), processes, people, places, systems and devices that can be used for various purposes. The digital representation provides both the elements and the dynamics of how an Internet of Things (IoT) device operates and lives throughout its life cycle. Definitions of digital twin technology used in prior research emphasize two important characteristics. Firstly, each definition emphasizes the connection between the physical model and the corresponding virtual model or virtual counterpart. Secondly, this connection is established by generating real-time data using sensors.[7] The concept of the digital twin can be compared to other concepts such as cross-reality environments or co-spaces and mirror models, which aim to, by and large, synchronize part of the physical world (e.g., an object or place) with its cyber representation (which can be an abstraction of some aspects of the physical world)[8].

Digital twins integrate IoT, artificial intelligence, machine learning and software analytics with spatial network graphs to create living digital simulation models that update and change as their physical counterparts change. A digital twin continuously learns and updates itself from multiple sources to represent its near real-time status, working condition or position. This learning system learns from itself, using sensor data that conveys various aspects of its operating condition; from human experts, such as engineers with deep and relevant industry domain knowledge; from other similar machines; from other similar fleets of machines; and from the larger systems and environment of which it may be a part. A digital twin also integrates historical data from past machine usage to factor into its digital model.

In various industrial sectors, twins are being used to optimize the operation and maintenance of physical assets, systems and manufacturing processes. They are a formative technology for the Industrial Internet of Things (IIoT), where physical objects can live and interact with other machines and people virtually. In the context of the IoT, they are also referred to as "cyber objects" or "digital avatars". The digital twin is also a component of cyber-physical systems.

13.2.2 Origin and Types of Digital Twins

Digital twins were anticipated by David Gelernter's 1991 book Mirror Worlds. It is widely acknowledged in both industry and academic publications that Michael Grieves of Florida Institute of Technology first applied the digital twin concept in manufacturing. The concept and model of the digital twin was publicly introduced in 2002 by Grieves, then of the University of Michigan, at a Society of Manufacturing Engineers conference in Troy, Michigan. Grieves proposed the digital twin as the conceptual model underlying product lifecycle management (PLM).

An Early Digital Twin Concept by Grieves and Vickers

The concept, which had a few different names, was subsequently called the "digital twin" by John Vickers of NASA in a 2010 Roadmap Report. The digital twin concept consists of three distinct parts: the physical product, the digital/virtual product, and connections between the two products. The connections between the physical product and the digital/virtual product are data that flows from the physical product to the digital/virtual product and information that is available from the digital/virtual product to the physical environment.

The concept was divided into types later. The types are the digital twin prototype (DTP), the digital twin instance (DTI), and the digital twin aggregate (DTA). The DTP consists of the designs, analyses, and processes to realize a physical product. The DTP exists before there is a physical product. The DTI is the digital twin of each individual instance of the product once it is manufactured. The DTA is the aggregation of DTIs whose data and information can be used for interrogation about the physical product, prognostics, and learning. The specific information contained in the digital twins is driven by use cases. The digital twin is a logical construct, meaning that the actual data and information may be contained in other applications.

A digital twin in the workplace is often considered part of robotic process automation (RPA) and, per industry-analyst firm Gartner, is part of the broader and emerging "hyper automation" category.

13. 2. 3 Application of Digital Twins

An example of how digital twins are used to optimize machines is with the maintenance of power generation equipment such as power generation turbines, jet engines and locomotives.

Another example of digital twins is the use of 3D modeling to create digital companions for the physical objects. It can be used to view the status of the actual physical object, which provides a way to project physical objects into the digital world. For example, when sensors collect data from a connected device, the sensor data can be used to update a "digital twin" copy of the device's state in real time. The term "device shadow" is also used for the concept of a digital twin. The digital twin is meant to be an up-to-date and accurate copy of the physical object's properties and states, including shape, position, gesture, status, and motion.

A digital twin also can be used for monitoring, diagnostics, and prognostics to optimize asset performance and utilization. In this field, sensory data can be combined with historical data, human expertise and fleet and simulation learning to improve the outcome of prognostics. Therefore, complex prognostics and intelligent maintenance system platforms can use digital twins in finding the root cause of issues and improve productivity.

Digital twins of autonomous vehicles and their sensor suite embedded in a traffic and environment simulation have also been proposed as a means to overcome the significant development, testing and validation challenges for the automotive application, in particular when the related algorithms are based on artificial intelligence approaches that require extensive training data and validation data sets.

Further examples of industry applications:

(1) Aircraft engines;

(2) Wind turbines;

(3) Large structures, e. g. , offshore platforms, offshore vessels etc. ;

(4) HVAC control systems;

(5) Locomotives;

(6) Buildings;

(7) Utilities (electric, gas, water, wastewater networks).

The physical manufacturing objects are virtualized and represented as digital twin models (avatars) seamlessly and closely integrated in both the physical and cyber spaces. Physical objects and twin models interact in a mutually beneficial manner.

13. 2. 3. 1 Industry-Level Dynamics

The digital twin is disrupting the entire product lifecycle management (PLM), from design, to manufacturing to service and operations. Nowadays, PLM is very time consuming in terms of efficiency, manufacturing, intelligence, service phases and sustainability in product design. A digital twin can merge the product physical and virtual space. The digital twin enables companies to have a digital footprint of all of their products, from design to development and throughout the entire product life cycle. Broadly speaking, industries with manufacturing business are highly disrupted by digital twins. In the manufacturing process, the digital twin is like a virtual replica of the near-time occurrences in the factory. Thousands of sensors are being placed throughout the physical manufacturing process, all collecting data from different dimensions, such as environmental conditions, behavioral characteristics of the machine and work that is being performed. All this data is continuously communicating and collected by the digital twin.

Due to the Internet of Things, digital twins have become more affordable and could drive the future of the manufacturing industry. A benefit for engineers lays in real-world usage of products that are virtually being designed by the digital twin. Advanced ways of product and asset maintenance and management come within reach as there is a digital twin of the real "thing" with real-time capabilities.

Digital twins offer a great amount of business potential by predicting the future instead of analyzing the past of the manufacturing process. The representation of reality created by digital twins allows manufacturers to evolve towards ex - ante business practices. The future of manufacturing drives on the following four aspects: modularity, autonomy, connectivity, and digital twin. As there is an increasing digitalization in the stages of a manufacturing process, opportunities are opening up to achieve a higher productivity. This starts with modularity and leading to higher effectiveness in the production system. Furthermore, autonomy enables the production system to respond to unexpected events in an efficient and intelligent way. Lastly, connectivity like the Internet of Things, makes the closing of the digitalization loop possible, by then allowing the following cycle of product design and promotion to be optimized for higher performance. This may lead to increase in customer satisfaction and loyalty when products can determine a problem before actually breaking down. Furthermore, as storage and computing costs are becoming less expensive, the ways in which digital twins are used are expanding.

13. 2. 3. 2 Embedded Digital Twin

Remembering that a definition of digital twin is a real time digital replica of a physical device, manufacturers are embedding digital twin in their device. Proven advantages are improved quality, earlier fault detection and better feedback on product usage to product designer.

13. 2. 4 The Characteristics of Digital Twins

Digital technologies have certain characteristics that distinguish them from other technologies. These characteristics, in turn, have certain consequences. Digital twins have the following characteristics.

13. 2. 4. 1 Connectivity

One of the main characteristics of digital twin technology is its connectivity. The recent development of the Internet of Things (IoT) brings forward numerous new technologies. The development of IoT also brings forward the development of digital twin technology. This technology shows many characteristics that have similarities with the character of the IoT, namely its connective nature. First and foremost, the technology enables connectivity between the physical component and its digital counterpart. The basis of digital twins is based on this connection, without it, digital twin technology would not exist. As described in the previous section, this connectivity is created by sensors on the physical product which obtain data and integrate and communicate this data through various integration technologies. Digital twin technology enables increased connectivity between organizations, products, and customers. For example, connectivity between partners in a supply chain can be increased by enabling members of this supply chain to check the digital twin of a product or asset. These partners can then check the status of this product by simply checking the digital twin.

Also, connectivity with customers can be increased.

Servitization is the process of organizations that are adding value to their core corporate offerings through services. In the case of the example of engines, the manufacturing of the engine is the core offering of this organization, they then add value by providing a service of checking the engine and offering maintenance.

13.2.4.2 Homogenization

Digital twins can be further characterized as a digital technology that is both the consequence and an enabler of the homogenization of data. Due to the fact that any type of information or content can now be stored and transmitted in the same digital form, it can be used to create a virtual representation of the product (in the form of a digital twin), thus decoupling the information from its physical form[9]. Therefore, the homogenization of data and the decoupling of the information from its physical artifact, have allowed digital twins to come into existence. However, digital twins also enable increasingly more information on physical products to be stored digitally and become decoupled from the product itself.

As data is increasingly digitized, it can be transmitted, stored, and computed in fast and low-cost ways. According to Moore's law, computing power will continue to increase exponentially over the coming years, while the cost of computing decreases significantly. This would, therefore, lead to lower marginal costs of developing digital twins and make it comparatively much cheaper to test, predict, and solve problems on virtual representations rather than testing on physical models and waiting for physical products to break before intervening.

Another consequence of the homogenization and decoupling of information is that the user experience converges. As information from physical objects is digitized, a single artifact can have multiple new affordances. Digital twin technology allows detailed information about a physical object to be shared with a larger number of agents, unconstrained by physical location or time. In his white paper on digital twin technology in the manufacturing industry, Michael Grieves noted the following about the consequences of homogenization enabled by digital twins.

In the past, factory managers had their office overlooking the factory so that they could get a feel for what was happening on the factory floor. With the digital twin, not only the factory manager, but everyone associated with factory production could have that same virtual window to not only a single factory, but to all the factories across the globe.

13.2.4.3 Reprogrammable and Smart

As stated above, a digital twin enables a physical product to be reprogrammable in a certain way. Furthermore, the digital twin is also reprogrammable in an automatic manner. Through the sensors on the physical product, artificial intelligence technologies, and predictive analytics, A consequence of this reprogrammable nature is the emergence of functionalities. If we take the example of an engine again, digital twins can be used to collect data about the performance of the engine and if needed adjust the engine, creating a newer version of the product. Also, servitization can be seen as a consequence of the reprogrammable nature as well. Manufactures can be responsible for observing the digital twin, adjusting, or reprogramming the digital twin when needed and they can offer this as an extra service.

13.2.4.4 Digital Traces

Another characteristic that can be observed, is the fact that digital twin technologies leave digital traces. These traces can be used by engineers for example, when a machine malfunctions to go back and check the traces of the digital twin, to diagnose where the problem occurred. These diagnoses can in the future also be used by the manufacturer of these machines, to improve their designs so that these same malfunctions will occur less often in the future.

13.2.4.5 Modularity

In the sense of the manufacturing industry, modularity can be described as the design and customization of products and production modules. By adding modularity to the manufacturing models, manufacturers gain the ability to tweak models and machines. Digital twin technology enables manufacturers to track the machines that are used and notice possible areas of improvement in the machines. When these machines are made modular, by using digital twin technology, manufacturers can see which components make the machine perform poorly and replace these with better fitting components to improve the manufacturing process.

13.3 Artificial Intelligence

13.3.1 Machine Learning

Machine learning (ML) is the study of computer algorithms that improve automatically through experience. It is seen as a subset of artificial intelligence. Machine learning algorithms build a model based on sample data, known as "training data", in order to make predictions or decisions without being explicitly programmed to do so. Machine learning algorithms are used in a wide variety of applications, such as email filtering and computer vision, where it is difficult or infeasible to develop conventional algorithms to perform the needed tasks.

A subset of machine learning is closely related to computational statistics, which focuses on making predictions using computers; but not all machine learning is statistical learning. The study of mathematical optimization delivers methods, theory, and application domains to the field of machine learning. Data mining is a related field of study, focusing on exploratory data analysis through unsupervised learning[10]. In its application across business problems, machine learning is also referred to as predictive analytics.

13. 3. 2 Artificial Neural Networks

Artificial neural networks (ANNs), usually simply called neural networks (NNs), are computing systems vaguely inspired by the biological neural networks that constitute animal brains.

An ANN is based on a collection of connected units or nodes called artificial neurons, which loosely model the neurons in a biological brain. Each connection, like the synapses in a biological brain, can transmit a signal to other neurons. An artificial neuron that receives a signal then processes it and can signal neurons connected to it. The "signal" at a connection is a real number, and the output of each neuron is computed by some non-linear function of the sum of its inputs. The connections are called edges. Neurons and edges typically have a weight that adjusts as learning proceeds. The weight increases or decreases the strength of the signal at a connection. Neurons may have a threshold such that a signal is sent only if the aggregate signal crosses that threshold. Typically, neurons are aggregated into layers. Different layers may perform different transformations on their inputs. Signals travel from the first layer (the input layer) to the last layer (the output layer), possibly after traversing the layers multiple times.

13. 3. 3 The Application Status of Artificial Intelligence

Artificial intelligence, defined as intelligence exhibited by machines, has many applications in today's society. More specifically, it is Weak AI, the form of AI where programs are developed to perform specific tasks, that is being utilized for a wide range of activities including medical diagnosis, electronic trading platforms, robot control, and remote sensing. AI has been used to develop and advance numerous fields and industries, including finance, healthcare, education, transportation, and more.

13. 3. 3. 1 Finance

Financial institutions have long used artificial neural network systems to detect charges or claims outside of the norm, flagging these for human investigation. The use of AI in banking can be traced back to 1987 when Security Pacific National Bank in the US set-up a Fraud Prevention Task force to counter the unauthorized use of debit cards . Programs like Kasisto and Moneystream are using AI in financial services.

Banks use artificial intelligence systems today to organize operations, maintain book-keeping, invest in stocks, and manage properties. AI can react to changes overnight or when business is not taking place. In August 2001, robots beat humans in a simulated financial trading competition. [15] AI has also reduced fraud and financial crimes by monitoring behavioral patterns of users for any abnormal changes or anomalies[11].

AI is increasingly being used by corporations. Jack Ma has controversially predicted that AI CEOs are 30 years away.

The use of AI machines in the market in applications such as online trading and decision making has changed major economic theories. For example, AI-based buying and selling platforms have changed the law of supply and demand in that it is now possible to easily estimate individualized demand and supply curves and thus individualized pricing. Furthermore, AI machines reduce information asymmetry in the market and thus making markets more efficient while reducing the volume of trades (citation needed). Furthermore, AI in the markets limits the consequences of behavior in the markets again making markets more efficient. Other theories where AI has had impact include in rational choice, rational expectations, game theory, Lewis turning point, portfolio optimization and counterfactual thinking. In August 2019, the AICPA introduced an AI training course for accounting professionals.

13.3.3.2 Education

AI tutors could allow for students to get extra, one-on-one help. They could also reduce anxiety and stress for some students, that may be caused by tutor labs or human tutors. In future classrooms, ambient informatics can play a beneficial role. Ambient informatics is the idea that information is everywhere in the environment and that technologies automatically adjust to your personal preferences. Study devices could be able to create lessons, problems, and games to tailor to the specific student's needs, and give immediate feedback.

But AI can also create a disadvantageous environment with revenge effects, if technology is inhibiting society from moving forward and causing negative, unintended effects on society. An example of a revenge effect is that the extended use of technology may hinder students' ability to focus and stay on task instead of helping them learn and grow. Also, AI has been known to lead to the loss of both human agency and simultaneity.

13.4 Big Data Processing

Smart manufacturing utilizes big data analytics, to refine complicated processes and manage supply chains. Big data analytics refers to a method for gathering and understanding large data sets in terms of what are known as the three V's, **velocity**, variety, and volume. Velocity informs the frequency of data acquisition, which can be concurrent with the application of previous data. Variety describes the different types of data that may be handled. Volume represents the amount of data. Big data analytics allows an enterprise to use smart manufacturing to predict demand and the need for design changes rather than reacting to orders placed. Figure 13-2 shows graphic of a sample manufacturing control system showing the interconnectivity of data analysis, computing and automation.

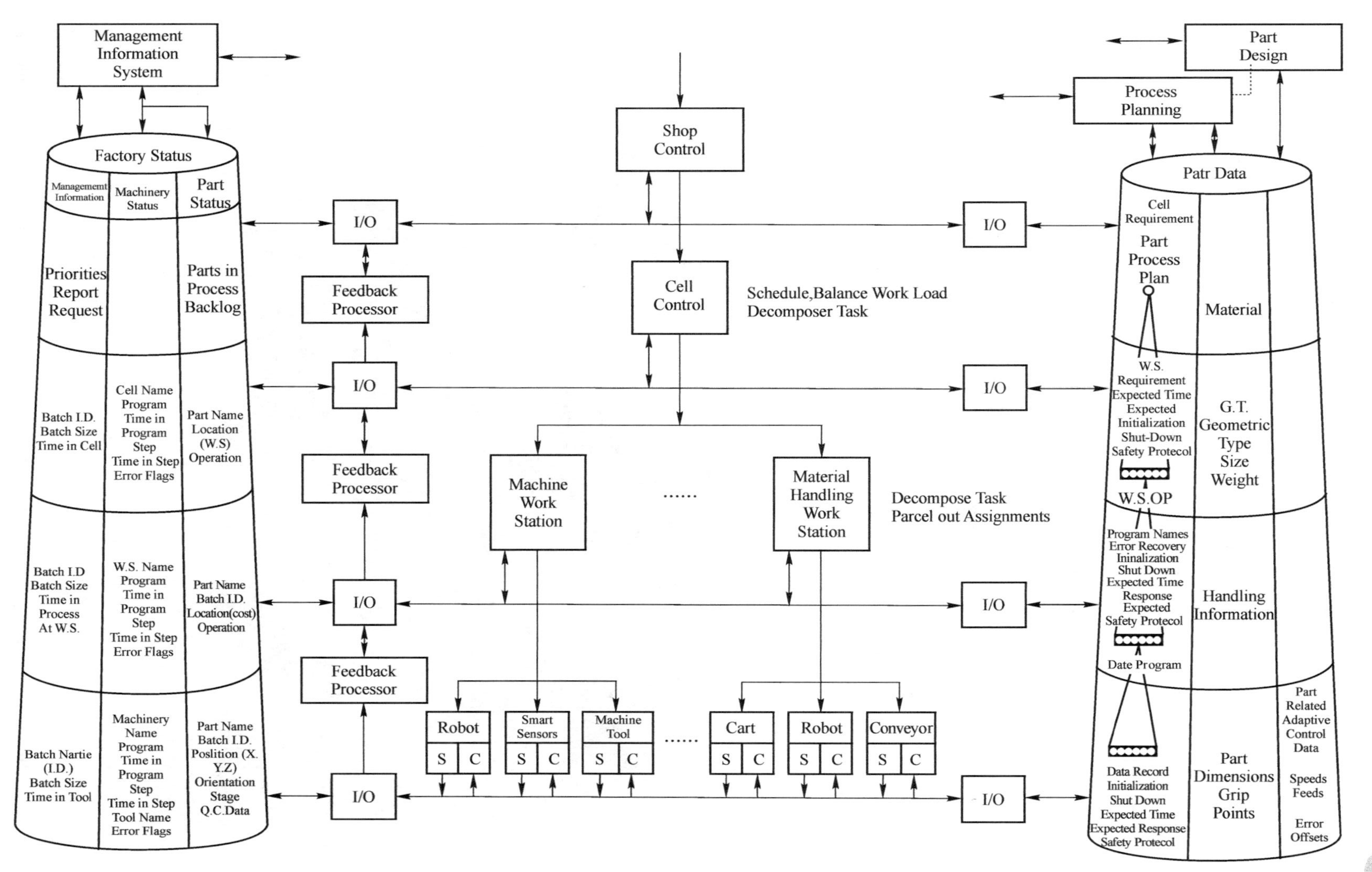

Figure 13-2 Graphic of a sample manufacturing control system showing the interconnectivity of data analysis, computing and automation

Some products have embedded sensors, which produce large amounts of data that can be used to understand consumer behavior and improve future versions of the product[12].

13.5 Advanced Industrial Robotics for Manufacturing

Advanced industrial robots, also known as smart machines operate autonomously and can communicate directly with manufacturing systems. In some advanced manufacturing contexts, they can work with humans for co-assembly tasks. By evaluating sensory input and distinguishing between different product configurations, these machines are able to solve problems and make decisions independent of people. These robots are able to complete work beyond what they were initially programmed to do and have artificial intelligence that allows them to learn from experience. These machines have the flexibility to be reconfigured and re-purposed. This gives them the ability to respond rapidly to design changes and innovation, which is a competitive advantage over more traditional manufacturing processes. An area of concern surrounding advanced robotics is the safety and well-being of the human workers who interact with robotic systems. Traditionally, measures have been taken to segregate robots from the human workforce but advances in robotic cognitive ability have opened up opportunities, such as robots, for robots to work collaboratively with people. Figure 13-3 shows advanced robotics used in automotive production.

扫一扫查看彩图

Figure 13-3 Advanced robotics used in automotive production

13.5.1 Types and Features of Industrial robot

(1) Articulated robots.

Articulated robots are the most common industrial robots. They look like a human arm, which is why they are also called robotic arm or manipulator arm. Their articulations with several degrees of freedom allow the articulated arms a wide range of movements.

(2) Cartesian coordinate robots.

Cartesian robots, also called rectilinear, gantry robots, and x-y-z robots have three prismatic joints for the movement of the tool and three rotary joints for its orientation in space.

To be able to move and orient the effector organ in all directions, such a robot needs 6 axes (or degrees of freedom). In a 2-dimensional environment, three axes are sufficient, two for displacement and one for orientation.

(3) Cylindrical coordinate robots.

The cylindrical coordinate robots are characterized by their rotary joint at the base and at least one prismatic joint connecting its links. They can move vertically and horizontally by sliding. The compact effector design allows the robot to reach tight workspaces without any loss of speed.

(4) Spherical coordinate robots.

Spherical coordinate robots only have rotary joints. They are one of the first robots to have been used in industrial applications. They are commonly used for machine tending in die-casting, plastic injection and extrusion, and for welding[13].

(5) SCARA robots.

SCARA is an acronym for Selective Compliance Assembly Robot Arm. SCARA robots are recognized by their two parallel joints which provide movement in the X-Y plane. Rotating shafts are positioned vertically at the effector.

SCARA robots are used for jobs that require precise lateral movements. They are ideal for assembly applications.

(6) Delta robots.

Delta robots are also referred to as parallel link robots. They consist of parallel links connected to a common base. Delta robots are particularly useful for direct control tasks and high maneuvering operations (such as quick pick-and-place tasks). Delta robots take advantage of four bar or parallelogram linkage systems.

13.5.2 History of Industrial Robotics

The earliest known industrial robot, conforming to the ISO definition was completed by "Bill" Griffith P. Taylor in 1937 and published in Meccano Magazine, March 1938. The crane-like device was built almost entirely using Meccano parts, and powered by a single electric motor. Five axes of movement were possible, including grab and grab rotation. Automation was achieved using punched paper tape to energize solenoids, which would facilitate the movement of the crane's control levers. The robot could stack wooden blocks in pre-programmed patterns. The number of motor revolutions required for each desired movement was first plotted on graph paper. This information was then transferred to the paper tape, which was also driven by the robot's single motor. Chris Shute built a complete replica of the robot in 1997.

George Devol applied for the first robotics patents in 1954 (granted in 1961). The first company to produce a robot was Unimation, founded by Devol and Joseph F. Engelberger in 1956. Unimation robots were also called programmable transfer machines since their main use at first was to transfer objects from one point to another, less than a dozen feet or so apart. They used hydraulic actuators and were programmed in joint coordinates, i. e. , the angles of the various joints were stored during a teaching phase and replayed in operation. They were accurate to within 1/10,000 of an inch (note: although accuracy is not an appropriate measure for robots, usually evaluated in terms of repeatability-see later). Unimation later licensed their technology to Kawasaki Heavy Industries and GKN, manufacturing Unimates in Japan and England respectively. For some time Unimation's only competitor was Cincinnati Milacron Inc. of Ohio. This changed radically in the late 1970s when several big Japanese conglomerates began producing similar industrial robots.

In 1969 Victor Scheinman at Stanford University invented the Stanford arm, an all-electric, 6-axis articulated robot designed to permit an arm solution. This allowed it accurately to follow arbitrary paths in space and widened the potential use of the robot to more sophisticated applications such as assembly and welding. Scheinman then designed a second arm for the MIT AI Lab, called the "MIT arm" . Scheinman, after receiving a fellowship from Unimation to develop his designs, sold those designs to Unimation who further developed them with support from General Motors and later marketed it as the Programmable Universal Machine for Assembly (PUMA).

Industrial robotics took off quite quickly in Europe, with both ABB Robotics and KUKA Robotics bringing robots to the market in 1973. ABB Robotics (formerly ASEA) introduced IRB 6, among the world's first commercially available all electric microprocessor-controlled robot. The first two IRB 6 robots were sold to Magnusson in Sweden for grinding and polishing pipe bends and were installed in production in January 1974. Also, in 1973 KUKA Robotics built its first robot, known as FAMULUS, also one of the first articulated robots to have six electromechanically driven axes.

Interest in robotics increased in the late 1970s and many US companies entered the field, including large firms like General Electric, and General Motors (which formed joint venture FANUC Robotics with FANUC LTD of Japan). U. S. startup companies included Automatic and Adept Technology, Inc. At the height of the robot boom in 1984, Unimation was acquired by Westinghouse Electric Corporation for 107 million U. S. dollars. Westinghouse sold Unimation to Stäubli Faverges SCA of France in 1988, which is still making articulated robots for general industrial and cleanroom applications and even bought the robotic division of Bosch in late 2004.

Only a few non-Japanese companies ultimately managed to survive in this market, the major ones being: Adept Technology, Stäubli, the Swedish-Swiss company ABB Asea Brown Boveri, the German company KUKA Robotics and the Italian company Comau.

13.5.3 Technical Description

13.5.3.1 Defining Parameters

(1) Number of axes—Two axes are required to reach any point in a plane; three axes are required to reach any point in space. To fully control the orientation of the end of the arm (i. e. , the wrist) three more axes (yaw, pitch, and roll) are required. Some designs (e. g. , he SCARA robot) trade limitations in motion possibilities for cost, speed, and accuracy.

(2) Degrees of freedom—This is usually the same as the number of axes.

(3) Working envelope—The region of space a robot can reach.

(4) Kinematics—The actual arrangement of rigid members and joints in the robot, which determines the robot's possible motions. Classes of robot kinematics include articulated, cartesian, parallel and SCARA.

(5) Carrying capacity or payload—How much weight a robot can lift.

(6) Speed—How fast the robot can position the end of its arm. This may be defined in terms of the angular or linear speed of each axis or as a compound speed i. e., the speed of the end of the arm when all axes are moving.

(7) Acceleration—How quickly an axis can accelerate. Since this is a limiting factor, a robot may not be able to reach its specified maximum speed for movements over a short distance or a complex path requiring frequent changes of direction.

(8) Accuracy—How closely a robot can reach a commanded position. When the absolute position of the robot is measured and compared to the commanded position the error is a measure of accuracy. Accuracy can be improved with external sensing for example a vision system or Infra-Red. See robot calibration. Accuracy can vary with speed and position within the working envelope and with payload (see compliance).

(9) Repeatability—How well the robot will return to a programmed position. This is not the same as accuracy. It may be that when told to go to a certain X-Y-Z position that it gets only to within 1 mm of that position. This would be its accuracy which may be improved by calibration. But if that position is taught into controller memory and each time it is sent there it returns to within 0.1mm of the taught position then the repeatability will be within 0.1mm.

Accuracy and repeatability are different measures. Repeatability is usually the most important criterion for a robot and is similar to the concept of 'precision' in measurement—see accuracy and precision. ISO 9283 sets out a method whereby both accuracy and repeatability can be measured. Typically, a robot is sent to a taught position a number of times and the error is measured at each return to the position after visiting 4 other positions. Repeatability is then quantified using the standard deviation of those samples in all three dimensions. A typical robot can, of course make a positional error exceeding that and that could be a problem for the process. Moreover, the repeatability is different in different parts of the working envelope and also changes with speed and payload. ISO 9283 specifies that accuracy and repeatability should be measured at maximum speed and at maximum payload. But this results in pessimistic values whereas the robot could be much more accurate and repeatable at light loads and speeds. Repeatability in an industrial process is also subject to the accuracy of the end effector, for example a gripper, and even to the design of the 'fingers' that match the gripper to the object being grasped. For example, if a robot picks a screw by its head, the screw could be at a random angle. A subsequent attempt to insert the screw into a hole could easily fail. These and similar scenarios can be improved with "lead-ins" e. g., by making the entrance to the hole tapered.

(1) Motion control—For some applications, such as simple pick-and-place assembly, the robot needs merely return repeatably to a limited number of pre-taught positions. For more sophisticated applications, such as welding and finishing (spray painting), motion must be

continuously controlled to follow a path in space, with controlled orientation and velocity.

(2) Power source—Some robots use electric motors; others use hydraulic actuators. The formers are faster, the latter are stronger and advantageous in applications such as spray painting, where a spark could set off an explosion; however, low internal air-pressurization of the arm can prevent ingress of flammable vapors as well as other contaminants. Nowadays, it is highly unlikely to see any hydraulic robots in the market. Additional sealings, brushless electric motors and spark-proof protection eased the construction of units that are able to work in the environment with an explosive atmosphere.

(3) Drive—Some robots connect electric motors to the joints via gears; others connect the motor to the joint directly (direct drive). Using gears results in measurable "backlash" which is free movement in an axis. Smaller robot arms frequently employ high speed, low torque DC motors, which generally require high gearing ratios; this has the disadvantage of backlash. In such cases the harmonic drive is often used.

(4) Compliance—This is a measure of the amount in angle or distance that a robot axis will move when a force is applied to it. Because of compliance when a robot goes to a position carrying its maximum payload it will be at a position slightly lower than when it is carrying no payload. Compliance can also be responsible for overshoot when carrying high payloads in which case acceleration would need to be reduced.

13.5.3.2 Robot Programming and Interfaces

Figure 13-4 shows a typical well-used teach pendant with optional mouse. The setup or programming of motions and sequences for an industrial robot is typically taught by linking the robot controller to a laptop, desktop computer or (internal or Internet) network.

扫一扫查看彩图

Figure 13-4 A typical well-used teach pendant with optional mouse

A robot and a collection of machines or peripherals is referred to as a work cell, or cell. A typical cell might contain a parts feeder, a molding machine, and a robot. The various machines are "integrated" and controlled by a single computer or PLC. How the robot interacts with other machines in the cell must be programmed, both with regard to their positions in the cell and synchronizing with them.

Software: The computer is installed with corresponding interface software. The use of a computer greatly simplifies the programming process. Specialized robot software is run either in the robot controller or in the computer or both depending on the system design.

There are two basic entities that need to be taught (or programmed): positional data and procedure. For example, in a task to move a screw from a feeder to a hole the positions of the feeder and the hole must first be taught or programmed. Secondly the procedure to get the screw from the feeder to the hole must be programmed along with any I/O involved, for example a signal to indicate when the screw is in the feeder ready to be picked up. The purpose of the robot software is to facilitate both these programming tasks.

Teaching the robot positions may be achieved a number of ways:

Positional commands the robot can be directed to the required position using a GUI or text based commands in which the required X-Y-Z position may be specified and edited.

Teach pendant: Robot positions can be taught via a teach pendant. This is a handheld control and programming unit. The common features of such units are the ability to manually send the robot to a desired position, or "inch" or "jog" to adjust a position. They also have a means to change the speed since a low speed is usually required for careful positioning, or while test-running through a new or modified routine. A large emergency stop button is usually included as well. Typically, once the robot has been programmed there is no more use for the teach pendant. All teach pendants are equipped with a 3-position Deadman switch. In the manual mode, it allows the robot to move only when it is in the middle position (partially pressed). If it is fully pressed in or completely released, the robot stops. This principle of operation allows natural reflexes to be used to increase safety.

Lead-by-the-nose: this is a technique offered by many robot manufacturers. In this method, one user holds the robot's manipulator, while another person enters a command which de-energizes the robot causing it to go into limp. The user then moves the robot by hand to the required positions and/or along a required path while the software logs these positions into memory. The program can later run the robot to these positions or along the taught path. This technique is popular for tasks such as paint spraying.

Offline programming is where the entire cell, the robot and all the machines or instruments in the workspace are mapped graphically. The robot can then be moved on screen and the process simulated. A robotics simulator is used to create embedded applications for a robot, without depending on the physical operation of the robot arm and end effector. The advantages of robotics simulation are that it saves time in the design of robotics applications. It can also increase the level of safety associated with robotic equipment since various "what if" scenarios can be tried and tested before the system is activated. Robot simulation software provides a platform to teach, test, run, and debug programs that have been written in a variety of programming languages. Figure 13-5 shows robotics simulator.

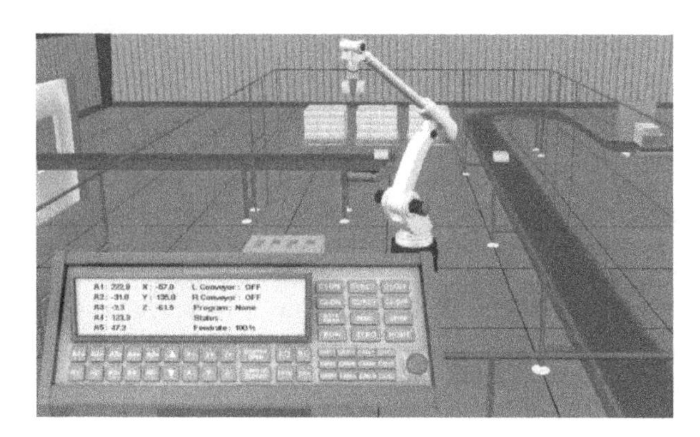

扫一扫查看彩图

Figure 13-5 Robotics simulator

Robot simulation tools allow for robotics programs to be conveniently written and debugged off-line with the final version of the program tested on an actual robot. The ability to preview the behavior of a robotic system in a virtual world allows for a variety of mechanisms, devices, configurations, and controllers to be tried and tested before being applied to a "real world" system. Robotics simulators have the ability to provide real-time computing of the simulated motion of an industrial robot using both geometric modeling and kinematics modeling.

Manufacturing independent robot programming tools are a relatively new but flexible way to program robot applications. Using a graphical user interface, the programming is done via drag and drop of predefined template/building blocks. They often feature the execution of simulations to evaluate the feasibility and offline programming in combination. If the system is able to compile and upload native robot code to the robot controller, the user no longer has to learn each manufacturer's proprietary language. Therefore, this approach can be an important step to standardize programming methods.

Other in addition, machine operators often use user interface devices, typically touchscreen units, which serve as the operator control panel. The operator can switch from program to program, adjust within a program and also operate a host of peripheral devices that may be integrated within the same robotic system. These include end effectors, feeders that supply components to the robot, conveyor belts, emergency stop controls, machine vision systems, safety interlock systems, barcode printers and an almost infinite array of other industrial devices which are accessed and controlled via the operator control panel.

The teach pendant or PC is usually disconnected after programming and the robot then runs on the program that has been installed in its controller. However, a computer is often used to 'supervise' the robot and any peripherals, or to provide additional storage for access to numerous complex paths and routines.

13.5.3.3 End-of-arm-tooling

The most essential robot peripheral is the end effector, or end-of-arm-tooling (EOT). Common examples of end effectors include welding devices (such as MIG-welding guns, spot-welders, etc.), spray guns and also grinding and deburring devices (such as pneumatic disk or belt grinders, burrs, etc.), and grippers (devices that can grasp an object, usually electromechanical or pneumatic). Other common means of picking up objects is by vacuum or magnets. End effectors are frequently highly complex, made to match the handled product and often capable of picking up an array of products at one time. They may utilize various sensors to aid the robot system in locating, handling, and positioning products.

13.5.3.4 Controlling Movement

For a given robot the only parameters necessary to completely locate the end effector (gripper, welding torch, etc.) of the robot are the angles of each of the joints or displacements of the linear axes (or combinations of the two for robot formats such as SCARA). However, there are many different ways to define the points. The most common and most convenient way of defining a point is to specify a Cartesian coordinate for it, i.e., the position of the 'end effector' in mm in the X, Y and Z directions relative to the robot's origin. In addition, depending on the types of joints a particular robot may have, the orientation of the end effector in yaw, pitch, and roll and the location of the tool point relative to the robot's faceplate must also be specified. For a jointed arm these coordinates must be converted to joint angles by the robot controller and such conversions are known as Cartesian Transformations which may need to be performed iteratively or recursively for a multiple axis robot. The mathematics of the relationship between joint angles and actual spatial coordinates is called kinematics.

Positioning by Cartesian coordinates may be done by entering the coordinates into the system or by using a teach pendant which moves the robot in X-Y-Z directions. It is much easier for a human operator to visualize motions up/down, left/right, etc. than to move each joint one at a time. When the desired position is reached it is then defined in some way particular to the robot software in use, e.g., P1 - P5 below.

13.5.3.5 Typical Programming

Most articulated robots perform by storing a series of positions in memory and moving to them at various times in their programming sequence. For example, a robot which is moving items from one place (bin A) to another (bin B) might have a simple "pick and place" program similar to the following:

Define points P1-P5:

Safely above workpiece (defined as P1)

10 cm Above bin A (defined as P2)

At position to take part from bin A (defined as P3)

10 cm Above bin B (defined as P4)

At position to take part from bin B. (defined as P5)

Define program:

Move to P1

Move to P2
Move to P3
Close gripper:
Move to P2
Move to P4
Move to P5
Open gripper:
Move to P4
Move to P1 and finish.

For examples of how this would look in popular robot languages see industrial robot programming.

13.5.3.6 Singularities

The American National Standard for Industrial Robots and Robot Systems — Safety Requirements (ANSI/RIA R15.06 - 1999) defines a singularity as "a condition caused by the collinear alignment of two or more robot axes resulting in unpredictable robot motion and velocities". It is most common in robot arms that utilize a "triple-roll wrist". This is a wrist about which the three axes of the wrist, controlling yaw, pitch, and roll, all pass through a common point. An example of a wrist singularity is when the path through which the robot is traveling causes the first and third axes of the robot's wrist (i.e., robot's axes 4 and 6) to line up. The second wrist axis then attempts to spin 180° in zero time to maintain the orientation of the end effector. Another common term for this singularity is a "wrist flip". The result of a singularity can be quite dramatic and can have adverse effects on the robot arm, the end effector, and the process. Some industrial robot manufacturers have attempted to side-step the situation by slightly altering the robot's path to prevent this condition. Another method is to slow the robot's travel speed, thus reducing the speed required for the wrist to make the transition. The ANSI/RIA has mandated that robot manufacturers shall make the user aware of singularities if they occur while the system is being manually manipulated.

A second type of singularity in wrist-partitioned vertically articulated six-axis robots occurs when the wrist center lies on a cylinder that is centered about axis 1 and with radius equal to the distance between axes 1 and 4. This is called a shoulder singularity. Some robot manufacturers also mention alignment singularities, where axes 1 and 6 become coincident. This is simply a sub-case of shoulder singularities. When the robot passes close to a shoulder singularity, joint 1 spin very fast.

The third and last type of singularity in wrist-partitioned vertically articulated six-axis robots occurs when the wrist's center lies in the same plane as axes 2 and 3.

Singularities are closely related to the phenomena of gimbal lock, which has a similar root cause of axes becoming lined up.

13. 6 Industrial Internet of Things

The Industrial Internet of Things (IIoT) refers to interconnected sensors, instruments, and other devices networked together with computers' industrial applications, including manufacturing and energy management. This connectivity allows for data collection, exchange, and analysis, potentially facilitating improvements in productivity and efficiency as well as other economic benefits. The IIoT is an evolution of a distributed control system (DCS) that allows for a higher degree of automation by using cloud computing to refine and optimize the process controls.

13. 6. 1 Overview of Industrial Internet of Things

The IIoT is enabled by technologies such as cybersecurity, cloud computing, edge computing, mobile technologies, machine-to-machine, 3D printing, advanced robotics, big data, internet of things, RFID technology, and cognitive computing[14]. Figure 13-6 shows IIOT architecture.

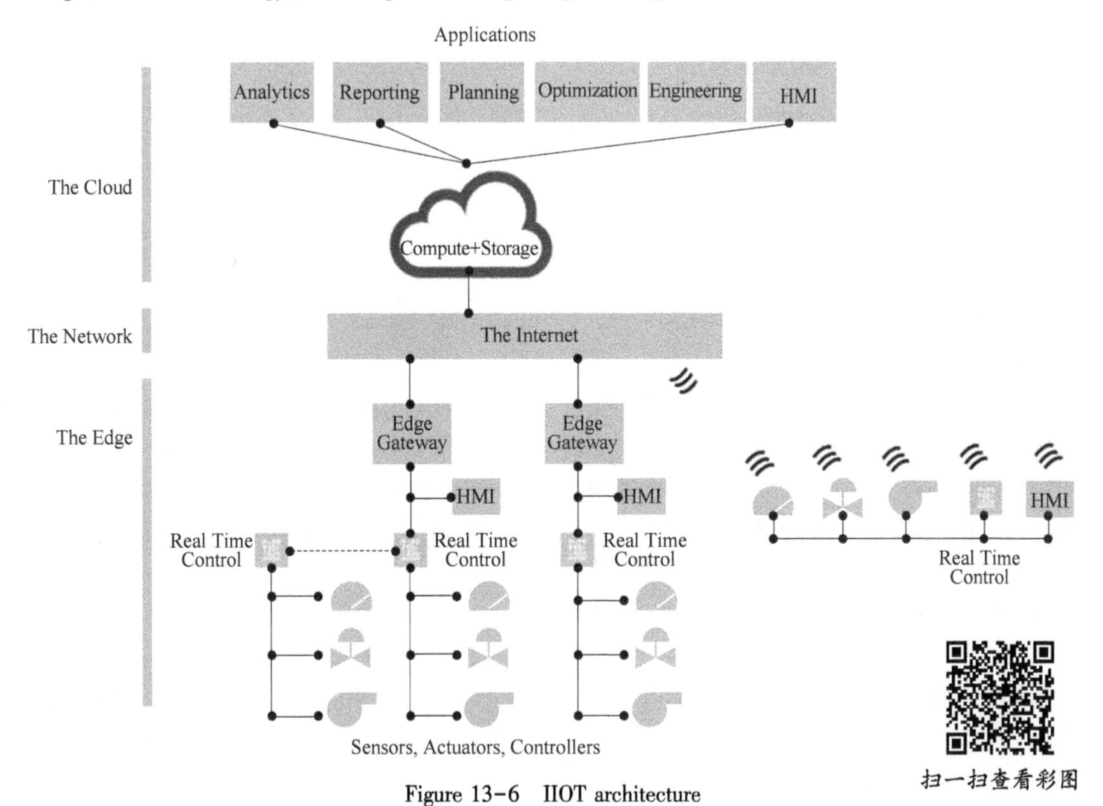

Figure 13-6 IIOT architecture

(1) Cyber-physical systems (CPS): the basic technology platform for IoT and IIoT and therefore the main enabler to connect physical machines that were previously disconnected. CPS integrates the dynamics of the physical process with those of software and communication, providing abstractions and modeling, design, and analysis techniques[15].

(2) Cloud computing: With cloud computing IT services and resources can be uploaded to and

retrieved from the Internet as opposed to direct connection to a server. Files can be kept on cloud-based storage systems rather than on local storage devices[16].

(3) Edge computing: A distributed computing paradigm which brings computer data storage closer to the location where it is needed. In contrast to cloud computing, edge computing refers to decentralized data processing at the edge of the network. The industrial internet requires more of an edge-plus-cloud architecture rather than one based on purely centralized cloud; in order to transform productivity, products, and services in the industrial world.

(4) Big data analytics: Big data analytics is the process of examining large and varied data sets, or big data.

(5) Artificial intelligence and machine learning: Artificial intelligence (AI) is a field within computer science in which intelligent machines are created that work and react like humans. Machine learning is a core part of AI, allowing software to predict outcomes more accurately without explicitly being programmed.

13.6.2 History of Industrial Internet of Things

The history of the IIoT begins with the invention of the programmable logic controller (PLC) by Dick Morley in 1968, which was used by General Motors in their automatic transmission manufacturing division. These PLCs allowed for fine control of individual elements in the manufacturing chain. In 1975, Honeywell and Yokogawa introduced the world's first DCSs, the TDC 2000 and the CENTUM system, respectively. These DCSs were the next step in allowing flexible process control throughout a plant, with the added benefit of backup redundancies by distributing control across the entire system, eliminating a singular point of failure in a central control room.

With the introduction of Ethernet in 1980, people began to explore the concept of a network of smart devices as early as 1982, when a modified Coke machine at Carnegie Mellon University became the first internet-connected appliance, able to report its inventory and whether newly loaded drinks were cold[17]. As early as in 1994, greater industrial applications were envisioned, as Reza Raji described the concept in IEEE Spectrum as "[moving] small packets of data to a large set of nodes, so as to integrate and automate everything from home appliances to entire factories" .

The concept of the internet of things first became popular in 1999, through the Auto-ID Center at MIT and related market-analysis publications. Radio-frequency identification (RFID) was seen by Kevin Ashton (one of the founders of the original Auto-ID Center) as a prerequisite for the internet of things at that point. If all objects and people in daily life were equipped with identifiers, computers could manage and inventory them. Besides using RFID, the tagging of things may be achieved through such technologies as near field communication, barcodes, QR codes and digital watermarking.

The current conception of the IIoT arose after the emergence of cloud technology in 2002, which allows for the storage of data to examine for historical trends, and the development of the OPC

Unified Architecture protocol in 2006, which enabled secure, remote communications between devices, programs, and data sources without the need for human intervention or interfaces.

One of the first consequences of implementing the industrial internet of things (by equipping objects with minuscule identifying devices or machine-readable identifiers) would be to create instant and ceaseless inventory control[18]. Another benefit of implementing an IIoT system is the ability to create a digital twin of the system. Using this digital twin allows for further optimization of the system by allowing for experimentation with new data from the cloud without having to halt production or sacrifice safety, as the new processes can be refined virtually until they are ready to be implemented. A digital twin can also serve as a training ground for new employees who won't have to worry about real impacts on the live system.

13.6.3 Application and Industries of Industrial Internet of Things

The term industrial internet of things is often encountered in the manufacturing industries, referring to the industrial subset of the IoT. Potential benefits of the industrial internet of things include improved productivity, analytics, and the transformation of the workplace. The potential of growth by implementing IIoT is predicted to generate $15 trillion of global GDP by 2030.

While connectivity and data acquisition are imperative for IIoT, they are not the end goals, but rather the foundation and path to something bigger. Of all the technologies, predictive maintenance is an "easier" application, as it is applicable to existing assets and management systems. Intelligent maintenance systems can reduce unexpected downtime and increase productivity, which is projected to save up to 12% over scheduled repairs, reduce overall maintenance costs up to 30%, and eliminate breakdowns up to 70%, according to some studies. Cyber-physical systems (CPS) are the core technology of industrial big data, and they will be an interface between human and the cyber world[19].

Integration of sensing and actuation systems connected to the Internet can optimize energy consumption as a whole. It is expected that IoT devices will be integrated into all forms of energy consuming devices (switches, power outlets, bulbs, televisions, etc.) and be able to communicate with the utility supply company in order to effectively balance power generation and energy usage. Besides home based energy management, the IIoT is especially relevant to the Smart Grid since it provides systems to gather and act on energy and power-related information in an automated fashion with the goal to improve the efficiency, reliability, economics, and sustainability of the production and distribution of electricity. Using advanced metering infrastructure (AMI) devices connected to the Internet backbone, electric utilities can not only collect data from end-user connections, but also manage other distribution automation devices like transformers and reclosers.

As of 2016, other real-world applications include incorporating smart LEDs to direct shoppers to empty parking spaces or highlight shifting traffic patterns, using of sensors on water purifiers to alert managers via computer or smartphone when to replace parts, attaching RFID tags to safety gear to track personnel and ensure their safety, embedding computers into power tools to record and track the torque level of individual tightening, and collecting data from multiple systems to enable the simulation of new processes.

13.6.3.1 Automotive Industry

Using IIoT in car manufacturing implies the digitalization of all elements of production. Software, machines, and humans are interconnected, enabling suppliers and manufacturers to rapidly respond to changing standards. IIoT enables efficient and cost-effective production by moving data from the customers to the company's systems, and then to individual sections of the production process. With IIoT, new tools and functionalities can be included in the manufacturing process. For example, 3D printers simplify the way of shaping pressing tools by printing the shape directly from steel granulate. These tools enable new possibilities for designing (with high precision). Customization of vehicles is also enabled by IIoT due to the modularity and connectivity of this technology. While in the past they worked separately, IIoT now enables humans and robots to cooperate. Robots take on the heavy and repetitive activities, so the manufacturing cycles are quicker, and the vehicle comes to the market more rapidly. Factories can quickly identify potential maintenance issues before they lead to downtime and many of them are moving to a 24-hour production plant, due to higher security and efficiency. The majority of automotive manufacturers companies have production plants in different countries, where different components of the same vehicle are built. IIoT makes possible to connect these production plants to each other, creating the possibility to move within facilities. Big data can be visually monitored which enables companies to respond faster to fluctuations in production and demand.

13.6.3.2 Oil and Gas Industry

With IIoT support, large amounts of raw data can be stored and sent by the drilling gear and research stations for cloud storage and analysis. With IIoT technologies, the oil and gas industry have the capability to connect machines, devices, sensors, and people through interconnectivity, which can help companies better address fluctuations in demand and pricing, address cybersecurity, and minimize environmental impact.

Across the supply chain, IIoT can improve the maintenance process, the overall safety, and connectivity. Drones can be used to detect possible oil and gas leaks at an early stage and at locations that are difficult to reach (e.g., offshore). They can also be used to identify weak spots in complex networks of pipelines with built-in thermal imaging systems. Increased connectivity (data integration and communication) can help companies with adjusting the production levels based on real-time data of inventory, storage, distribution pace, and forecasted demand. For example, a Deloitte report states that by implementing an IIoT solution integrating data from multiple internal and external sources (such as work management system, control center, pipeline attributes, risk scores, inline inspection findings, planned assessments, and leak history), thousands of miles of pipes can be monitored in real-time. This allows monitoring of pipeline threats, improving risk management, and providing situational awareness[20].

Benefits also apply to specific processes of the oil and gas industry. The exploration process of oil and gas can be done more precisely with 4D models built by seismic imaging[21]. These models map fluctuations in oil reserves and gas levels, they strive to point out the exact quantity of resources needed, and they forecast the lifespan of wells. The application of smart sensors and automated drillers gives companies the opportunity to monitor and produce more efficiently. Further, the storing process can also be improved with the implementation of IIoT by collecting and analyzing real-time data to monitor inventory levels and temperature control. IIoT can enhance the transportation process of oil and gas by implementing smart sensors and thermal detectors to give real-time geolocation data and monitor the products for safety reasons. These smart sensors can monitor the refinery processes and enhance safety. The demand for products can be forecasted more precisely and automatically be communicated to the refineries and production plants to adjust production levels.

13. 6. 3. 3 Agriculture Industry

In the agriculture industry, IIoT helps farmers to make decisions about when to harvest. Sensors collect data about soil and weather conditions and propose schedules for fertilizing and irrigating. Some livestock farms implant microchips into animals. This allows the farmers not only to trace their animals, but also pull up information about the lineage, weight, or health.

词汇

生词	音标	释义
explosive	[ɪkˈspləʊsɪv]	*adj.* 易爆炸的；可能引起爆炸的；易爆发的；可能引起冲动的；爆炸性的；暴躁的 *n.* 炸药；爆炸物
incorporates	[ɪnˈkɔːpəreɪts]	*n.* 将……包括在内；包含；吸收；使并入；注册成立
computerization	[kəmˌpjuːtəraɪˈzeɪʃn]	*v.* 将……包括在内；包含；吸收；使并入；注册成立
paradigms	[ˈpærədaɪmz]	*n.* 典范；范例；样式；词形变化表
parallel	[ˈpærəlel]	*adj.* 平行的；极相似的；同时发生的；相应的；对应的；并行的 *n.*（尤指不同地点或时间的）极其相似的人（或情况、事件等）；相似特征；相似特点；（地球或地图的）纬线，纬圈 *v.* 与……相似；与……同时发生；与……媲美；比得上
inherent	[ɪnˈherənt]	*adj.* 固有的；内在的

velocity	[vəˈlɒsətɪ]	*n.* （沿某一方向的）速度；高速；快速
cognitive	[ˈkɒgnətɪv]	*adj.* 认知的；感知的；认识的
synchronize	[ˈsɪŋkrəˌnaɪz]	*n.* 同步；手机同步；使同步；同步反应
explicitly	[ɪkˈsplɪsɪtlɪ]	*adv.* 明确地；明白地

长难句

In particular, new-generation intelligent manufacturing, which serves as the core technology of the current industrial revolution, incorporates major and profound changes in the development philosophy, manufacturing modes, and other aspects of the manufacturing industry.

尤其是，新一代智能制造作为当前工业革命的核心技术，将发展理念、制造模式和制造业其他方面融合在一起，并使其发生重大而深刻的变化。

Reference

[1] Rachuri, Sudarsan. Smart manufacturing systems design and analysis [J]. National Institute of Standards and Technology, 2014.

[2] Thareja, Priyavrat. A total quality organisation through people: (Part 4), building a world class foundry [J]. Foundry, 2006, XVIII (106): 37-43.

[3] Thareja, Priyavrat. Manufacturing paradigms in 2010 [R]. Radaur: National Conference on Emerging trends in Manufacturing Systems, JMIT, 2005.

[4] Wang W, Li R, Chen Y, et al. Facilitating human-robot collaborative tasks by teaching-learning-collaboration from human demonstrations [J]. IEEE Transactions on Automation Science and Engineering, 2019, 16 (2): 640-653.

[5] Saddik A El. Digital Twins: The Convergence of Multimedia Technologies [J]. IEEE Multi Media, 2019, 25 (2): 87-92.

[6] Loke S W, Thai B S, Torabi T, et al. The La trobe e-sanctuary: Building a cross-reality wildlife sanctuary [J]. 2015 International Conference on Intelligent Environments, 2015: 168-171.

[7] Gautier, Philippe. L'Internet des Objets... Internet, mais en mieux [M]. France: AFNOR, 2011.

[8] Mitchell T . Machine learning [M]. New York: McGraw Hill, 1997.

[9] Bennett F H, Andre D, Keane M A. Automated design of both the topology and sizing of analog electrical circuits using genetic programming [J]. Artificial Intelligence in Design '96. Springer, Dordrecht, 1996: 151-170.

[10] Christopher B. Pattern recognition and machine learning [M], Springer, 2006.

[11] CTO Corner: Artificial intelligence use in financial services-financial services roundtable [R]. Financial Services Roundtable, 2015.

[12] Marwala, Tshilidzi, Hurwitz, et al. Artificial intelligence and economic theory: Skynet in the market [M]. London: Springer, 2017.

[13] Anabel Q H. Technology and society: social networks, power, and inequality [M]. 2nd ed. London: Oxford University Press, 2016.

[14] Leveling J, Edelbrock M, Otto B. Big data analytics for supply chain management [J]. 2014 IEEE International Conference on Industrial Engineering and Engineering Management (IEEM), 2014: 918-922.

[15] Yang C, Shen W M, Wang X B. The internet of things in manufacturing: Key issues and potential applications [J]. IEEE Systems, Man, and Cybernetics Magazine, 2018, 4 (1): 6-15.

[16] Porter M E. How smart, connected products are transforming competition [M]. Boston: Harvard Business Review, 2016.

[17] James S A. English: architecture for the NBS automated manufacturing research facility (AMRF) [J]. National Institute of Standards and Technology, Gaithersburg, Maryland, 2016.

[18] Wang W, Li R, Chen Y, et al. Facilitating human-robot collaborative tasks by teaching-learning-collaboration from human demonstrations [J]. IEEE Transactions on Automation Science and Engineering, 2019, 16 (2): 640-653.

[19] Bicchi A, Peshkin M A, Colgate J. et al. Safety for physical human-robot interaction [J]. Springer Berlin Heidelberg, 2008: 1335-1348.

[20] Meccano. An automatic block-setting crane [N]. Meccano Magazine. 1938-03.

[21] Taylor G P. The robot gargantua [J]. Gargantua: Constructor Quarterly, 1995.